Cambridge International AS & A Level

Further Mathematics
Further Probability & Statistics

John du Feu
Series editor: Roger Porkess

Questions from the Cambridge International AS & A Level Further Mathematics papers are reproduced by permission of Cambridge Assessment International Education. Unless otherwise acknowledged, the questions, example answers, and comments that appear in this book were written by the authors. Cambridge Assessment International Education bears no responsibility for the example answers to questions taken from its past question papers which are contained in this publication.

*IGCSE is a registered trademark.

The publishers would like to thank the following who have given permission to reproduce photographs in this book:

Photo credits

p.1 © jakapan Chumchuen/Shutterstock; **p.10** © lcswart/Shutterstock; **p.28** © Barsan ATTILA/ Shutterstock; **p.78** © Dean Drobot/Shutterstock; **p.121** © Arthur Simoes/Shutterstock; **p.156** © Vibeke Morfield/123RF.

Hachette UK's policy is to use papers that are natural, renewable and recyclable products and made from wood grown in sustainable forests. The logging and manufacturing processes are expected to conform to the environmental regulations of the country of origin.

Orders: please contact Bookpoint Ltd, 130 Park Drive, Milton Park, Abingdon, Oxon OX14 4SE. Telephone: (44) 01235 827720. Fax: (44) 01235 400401. Email: education@bookpoint.co.uk. Lines are open from 9 a.m. to 5 p.m., Monday to Saturday, with a 24-hour message answering service. You can also order though our website: www.hoddereducation.com

Much of the material in this book was published originally as part of the MEI Structured Mathematics series. It has been carefully adapted for the Cambridge International AS & A Level Further Mathematics syllabus. The original MEI author team for Statistics comprised Alec Cryer, Michael Davies, Anthony Eccles, Bob Francis, Gerald Goodall, Alan Graham, Nigel Green, Liam Hennessy, Roger Porkess and Charlie Stripp.

© Roger Porkess and John du Feu 2018

First published in 2018 by
Hodder Education, an Hachette UK company,
Carmelite House, 50 Victoria Embankment,
London EC4Y 0DZ

Impression number 5 4 3 2 1

Year   2022   2021   2020   2019   2018

Cover photo by Shutterstock/Ed Samuel
Illustrations by Tech-Set, Aptara, Inc., and Integra Software Services
Typeset in Bembo Std 11/13 Integra Software Services Pvt Ltd, Pondicherry, India
Printed in Italy

A catalogue record for this title is available from the British Library.

ISBN 9781510421813

Contents

Introduction

This is one of a series of four books supporting the Cambridge International AS & A Level Further Mathematics 9231 syllabus for examination from 2020. It is preceded by five books supporting Cambridge International AS & A Level Mathematics 9709. The five chapters in this book cover the further probability and statistics required for the Paper 4 examination. This part of the series also contains two books for further pure mathematics and one book for further mechanics.

These books are based on the highly successful series for the Mathematics in Education and Industry (MEI) syllabus in the UK but they have been redesigned and revised for Cambridge International students; where appropriate, new material has been written and the exercises contain many past Cambridge International examination questions. An overview of the units making up the Cambridge International syllabus is given in the following pages.

Throughout the series, the emphasis is on understanding the mathematics as well as routine calculations. The various exercises provide plenty of scope for practising basic techniques; they also contain many typical examination-style questions.

The original MEI author team would like to thank John du Feu who has carried out the extensive task of presenting their work in a suitable form for Cambridge International students and for his many original contributions. They would also like to thank Cambridge Assessment International Education for its detailed advice in preparing the books and for permission to use many past examination questions.

Roger Porkess

Series editor

How to use this book

The structure of the book

This book has been endorsed by Cambridge Assessment International Education. It is listed as an endorsed textbook for students taking the Cambridge International AS & A Level Further Mathematics 9231 syllabus. The Further Probability & Statistics syllabus content is covered comprehensively and is presented across five chapters, offering a structured route through the course.

The book is written on the assumption that you have covered and understood the work in the Cambridge International AS & A Level Mathematics 9709 syllabus, including the probability and statistics content. The following icon is used to indicate material that is not directly on the syllabus.

> **e**   There are places where the book goes beyond the requirements of the syllabus to show how the ideas can be taken further or where fundamental underpinning work is explored. Such work is marked as **extension**.

Each chapter is broken down into several sections, with each section covering a single topic. Topics are introduced through **explanations**, with **key terms** picked out in red. These are reinforced with plentiful **worked examples**, punctuated with commentary, to demonstrate methods and illustrate application of the mathematics under discussion.

Regular **exercises** allow you to apply what you have learned. They offer a large variety of practice and higher-order question types that map to the key concepts of the Cambridge International syllabus. Look out for the following icons.

PS **Problem-solving questions** will help you to develop the ability to analyse problems, recognise how to represent different situations mathematically, identify and interpret relevant information, and select appropriate methods.

M **Modelling questions** provide you with an introduction to the important skill of mathematical modelling. In this, you take an everyday or workplace situation, or one that arises in your other subjects, and present it in a form that allows you to apply mathematics to it.

CP **Communication and proof questions** encourage you to become a more fluent mathematician, giving you scope to communicate your work with clear, logical arguments and to justify your results.

Exercises also include questions from real Cambridge Assessment International Education past papers, so that you can become familiar with the types of questions you are likely to meet in formal assessments.

Answers to exercise questions, excluding long explanations and proofs, are available online at www.hoddereducation.com/cambridgeextras, so you can check your work. It is important, however, that you have a go at answering the questions before looking up the answers if you are to understand the mathematics fully.

In addition to the exercises, a range of additional features are included to enhance your learning.

> ► **ACTIVITY**
>
> **Activities** invite you to do some work for yourself, typically to introduce you to ideas that are then going to be taken further. In some places, activities are also used to follow up work that has just been covered.

INVESTIGATION

In real life, it is often the case that as well as analysing a situation or problem, you also need to carry out some investigative work. This allows you to check whether your proposed approach is likely to be fruitful or to work at all, and whether it can be extended. Such opportunities are marked as **investigations**.

Other helpful features include the following.

 This symbol highlights points it will benefit you to **discuss** with your teacher or fellow students, to encourage deeper exploration and mathematical communication. If you are working on your own, there are answers available online at www.hoddereducation.com/cambridgeextras.

 This is a **warning** sign. It is used where a common mistake, misunderstanding or tricky point is being described to prevent you from making the same error.

A variety of notes are included to offer advice or spark your interest:

> ◤ Note
> ---
> **Notes** expand on the topic under consideration and explore the deeper lessons that emerge from what has just been done.

> ◤ Historical note
> ---
> **Historical notes** offer interesting background information about famous mathematicians or results to engage you in this fascinating field.

Technology note

Although graphical calculators and computers are not permitted in the examinations for this Cambridge International syllabus, we have included **Technology notes** to indicate places where working with them can be helpful for learning and for teaching.

Finally, each chapter ends with the **key points** covered, plus a list of the **learning outcomes** that summarise what you have learned in a form that is closely related to the syllabus.

Digital support

Comprehensive online support for this book, including further questions, is available by subscription to MEI's Integral® online teaching and learning platform for AS & A Level Mathematics and Further Mathematics, integralmaths.org. This online platform provides extensive, high-quality resources, including printable materials, innovative interactive activities, and formative and summative assessments. Our eTextbooks link seamlessly with Integral, allowing you to move with ease between corresponding topics in the eTextbooks and Integral.

MEI's Integral® material has not been through the Cambridge International endorsement process.

The Cambridge International AS & A Level Further Mathematics 9231 syllabus

The syllabus content is assessed over four examination papers.

Paper 1: Further Pure Mathematics 1	Paper 3: Further Mechanics
• 2 hours	• 1 hour 30 minutes
• 60% of the AS Level; 30% of the A Level	• 40% of the AS Level; 20% of the A Level
• Compulsory for AS and A Level	• Offered as part of AS; compulsory for A Level
Paper 2: Further Pure Mathematics 2	**Paper 4: Further Probability & Statistics**
• 2 hours	• 1 hour 30 minutes
• 30% of the A Level	• 40% of the AS Level; 20% of the A Level
• Compulsory for A Level; not a route to AS Level	• Offered as part of AS; compulsory for A Level

The following diagram illustrates the permitted combinations for AS Level and A Level.

AS Level Further Mathematics

A Level Further Mathematics

Paper 1 and Paper 3
Further Pure Mathematics 1
and Further Mechanics

Paper 1, 2, 3 and 4
Further Pure Mathematics 1 and 2,
Further Mechanics and Further
Probability & Statistics

Paper 1 and Paper 4
Further Pure Mathematics 1
and Further Probability & Statistics

Prior knowledge

It is expected that learners will have studied the majority of the Cambridge International AS & A Level Mathematics 9709 syllabus content before studying Cambridge International AS & A Level Further Mathematics 9231.

The prior knowledge required for each Further Mathematics component is shown in the following table.

Component in AS & A Level Further Mathematics 9231	Prior knowledge required from AS & A Level Mathematics 9709
9231 Paper 1: Further Pure Mathematics 1	9709 Papers 1 and 3
9231 Paper 2: Further Pure Mathematics 2	9709 Papers 1 and 3
9231 Paper 3: Further Mechanics	9709 Papers 1, 3 and 4
9231 Paper 4: Further Probability & Statistics	9709 Papers 1, 3, 5 and 6

For Paper 4: Further Probability & Statistics, knowledge of Cambridge International AS & A Level Mathematics 9709 Papers 5 and 6: Probability & Statistics subject content is assumed.

Command words

The table below includes command words used in the assessment for this syllabus. The use of the command word will relate to the subject context.

Command word	What it means
Calculate	work out from given facts, figures or information
Deduce	conclude from available information
Derive	obtain something (expression/equation/value) from another by a sequence of logical steps
Describe	state the points of a topic / give characteristics and main features
Determine	establish with certainty
Evaluate	judge or calculate the quality, importance, amount, or value of something
Explain	set out purposes or reasons / make the relationships between things evident / provide why and/or how and support with relevant evidence
Identify	name/select/recognise
Interpret	identify meaning or significance in relation to the context
Justify	support a case with evidence/argument
Prove	confirm the truth of the given statement using a chain of logical mathematical reasoning
Show (that)	provide structured evidence that leads to a given result
Sketch	make a simple freehand drawing showing the key features
State	express in clear terms
Verify	confirm a given statement/result is true

Key concepts

Key concepts are essential ideas that help students develop a deep understanding of mathematics.

The key concepts are:

Problem solving

Mathematics is fundamentally problem solving and representing systems and models in different ways. These include:

» Algebra: this is an essential tool which supports and expresses mathematical reasoning and provides a means to generalise across a number of contexts.

» Geometrical techniques: algebraic representations also describe a spatial relationship, which gives us a new way to understand a situation.

» Calculus: this is a fundamental element which describes change in dynamic situations and underlines the links between functions and graphs.

» Mechanical models: these explain and predict how particles and objects move or remain stable under the influence of forces.

» Statistical methods: these are used to quantify and model aspects of the world around us. Probability theory predicts how chance events might proceed, and whether assumptions about chance are justified by evidence.

Communication

Mathematical proof and reasoning is expressed using algebra and notation so that others can follow each line of reasoning and confirm its completeness and accuracy. Mathematical notation is universal. Each solution is structured, but proof and problem solving also invite creative and original thinking.

Mathematical modelling

Mathematical modelling can be applied to many different situations and problems, leading to predictions and solutions. A variety of mathematical content areas and techniques may be required to create the model. Once the model has been created and applied, the results can be interpreted to give predictions and information about the real world.

These key concepts are reinforced in the different question types included in this book: **Problem-solving**, **Communication and proof**, and **Modelling**.

1

Continuous random variables

> The control of large numbers is possible, and like unto that of small numbers, if we subdivide them.
>
> *Sun Tzu, 'The Art Of War' (544BC–496BC)*

You will recall having met **probability density functions (PDFs)** for continuous random variables in A Level Mathematics. To find probabilities using a probability density function f(x) you need to integrate the function between the limits you are using.

$$\text{So } P(a \leqslant X \leqslant b) = \int_a^b f(x)\,dx$$

Here is a summary of what you should be able to do:

» use a simple continuous random variable as a model

» understand the meaning of a probability density function (PDF) and be able to use one to find probabilities

» know and use the properties of a PDF

» sketch the graph of a PDF

» find the:
  - mean
  - variance
  - median
  - percentiles
  - mode

  from a given PDF

» use probability density functions to solve problems.

Answers to exercises are available at www.hoddereducation.com/cambridgeextras

Left margin vertical text

1.1 Piecewise definition of a probability density function

The definitions of the probability density functions that you have met so far only involve single functions. However, probability density functions can have two or more separate parts. In this case, the PDF is said to be defined piecewise. The following example illustrates this.

Example 1.1

The number of hours Nabilah spends each day working in her garden is modelled by the continuous random variable X, with PDF $f(x)$ defined by

$$f(x) = \begin{cases} kx & \text{for } 0 \leqslant x < 3 \\ k(6-x) & \text{for } 3 \leqslant x \leqslant 6 \\ 0 & \text{otherwise.} \end{cases}$$

(i) Find the value of k.

(ii) Sketch the graph of $f(x)$.

(iii) Find the probability that Nabilah will work between 2 and 5 hours in her garden on a randomly selected day.

Solution

(i) To find the value of k you must use the fact that the area under the graph of $f(x)$ is equal to 1. You may find the area by integration, as shown below.

$$\int_{-\infty}^{\infty} f(x)\,dx = \int_{0}^{3} kx\,dx + \int_{3}^{6} k(6-x)\,dx = 1$$

$$\left[\frac{kx^2}{2}\right]_0^3 + \left[6kx - \frac{kx^2}{2}\right]_3^6 = 1$$

Therefore $\quad \dfrac{9k}{2} + (36k - 18k) - \left(18k - \dfrac{9k}{2}\right) = 1$

$$9k = 1$$

So $\quad k = \dfrac{1}{9}$

(ii)

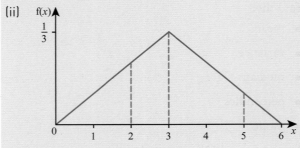

▲ Figure 1.1

> **Note**
>
> In this case you could have found k without integration because the graph of the PDF is a triangle, with area given by $\frac{1}{2} \times$ base $\times$ height, resulting in the equation
>
> $$\frac{1}{2} \times 6 \times k(6-3) = 1$$
>
> hence $\qquad\qquad 9k = 1$
>
> and $\qquad\qquad k = \frac{1}{9}.$

(iii) To find $P(2 \leqslant X \leqslant 5)$, you need to find both $P(2 \leqslant X < 3)$ and $P(3 \leqslant X \leqslant 5)$ because there is a different expression for each part.

$$P(2 \leqslant X \leqslant 5) = P(2 \leqslant X < 3) + P(3 \leqslant X \leqslant 5)$$

$$= \int_2^3 \frac{1}{9} x \, dx + \int_3^5 \frac{1}{9}(6-x) \, dx$$

$$= \left[\frac{x^2}{18}\right]_2^3 + \left[\frac{2x}{3} - \frac{x^2}{18}\right]_3^5$$

$$= \frac{9}{18} - \frac{4}{18} + \left(\frac{10}{3} - \frac{25}{18}\right) - \left(2 - \frac{1}{2}\right)$$

$$= 0.72 \text{ to two decimal places.}$$

The probability that Nabilah works between 2 and 5 hours in her garden on a randomly selected day is 0.72.

1.2 The expectation and variance of a function of X

There are times when one random variable is a function of another random variable. For example:

▸▸ as part of an experiment you are measuring temperatures in Celsius but then need to convert them to Fahrenheit: $F = 1.8C + 32$

▸▸ you are measuring the lengths of the sides of square pieces of material and deducing their areas: $A = L^2$

▸▸ you are estimating the ages, A years, of hedgerows by counting the number, n, of types of shrubs and trees in $30\,\text{m}$ lengths: $A = 100n - 50$.

In fact, in any situation where you are entering the value of a random variable into a formula, the outcome will be another random variable that is a function of the one you entered. Under these circumstances you may need to find the expectation and variance of such a function of a random variable.

Answers to exercises are available at www.hoddereducation.com/cambridgeextras

For a discrete random variable, X, in which the value x_i occurs with probability p_i, the expectation and variance of a function $g(X)$ are given by

$$E(g(X)) = \Sigma g(x_i) p_i$$

$$Var(g(X)) = \Sigma (g(x_i))^2 p_i - \{E(g(X))\}^2$$

The equivalent results for a continuous random variable, X, with PDF $f(x)$ are

$$E(g(X)) = \int_{\substack{\text{All} \\ \text{values} \\ \text{of } x}} g(x)f(x)\,dx$$

$$Var(g(X)) = \int_{\substack{\text{All} \\ \text{values} \\ \text{of } x}} (g(x))^2 f(x)\,dx - \{E(g(X))\}^2$$

You may find it helpful to think of the function $g(X)$ as a new variable; say, Y.

Example 1.2

The continuous random variable X has PDF $f(x)$ where

$$f(x) = \begin{cases} kx & \text{for } 0 \leqslant x \leqslant 2 \\ 4k - kx & \text{for } 2 < x \leqslant 4 \\ 0 & \text{otherwise.} \end{cases}$$

(i) Find the value of the constant k.

(ii) Sketch $y = f(x)$.

(iii) Find $P(1 \leqslant X \leqslant 3.5)$.

The continuous random variable $Y = X^2$.

(iv) Find $E(Y)$.

Solution

(i)
$$\int_0^2 kx\,dx + \int_2^4 (4k - kx)\,dx = 1$$

$$k\left[\frac{x^2}{2}\right]_0^2 + k\left[4x - \frac{x^2}{2}\right]_2^4 = 1$$

$$k(2 - 0) + k[(16 - 8) - (8 - 2)] = 1$$

$$2k + 2k = 1$$

$$k = \frac{1}{4}$$

(ii)

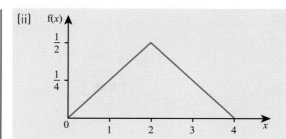

▲ **Figure 1.2**

(iii) $\mathrm{P}(1 \leqslant X \leqslant 3.5)$

$$= \int_1^2 \frac{1}{4}x \, \mathrm{d}x + \int_2^{3.5} \left(1 - \frac{1}{4}x\right)\mathrm{d}x$$

$$= \left[\frac{x^2}{8}\right]_1^2 + \left[x - \frac{x^2}{8}\right]_2^{3.5}$$

$$= \left(\frac{1}{2} - \frac{1}{8}\right) + \left[\left(3.5 - \frac{12.25}{8}\right) - \left(2 - \frac{1}{2}\right)\right]$$

$$= \frac{27}{32} = 0.84375$$

(iv) $\mathrm{E}(Y) = \int_0^2 x^2 \frac{1}{4}x \, \mathrm{d}x + \int_2^4 x^2 \left(1 - \frac{1}{4}x\right)\mathrm{d}x$

$$= \int_0^2 \frac{1}{4}x^3 \, \mathrm{d}x + \int_2^4 \left(x^2 - \frac{1}{4}x^3\right)\mathrm{d}x$$

$$= \left[\frac{x^4}{16}\right]_0^2 + \left[\frac{x^3}{3} - \frac{x^4}{16}\right]_2^4$$

$$= (1 - 0) + \left[\left(\frac{64}{3} - 16\right) - \left(\frac{8}{3} - 1\right)\right]$$

$$= \frac{14}{3}$$

Example 1.3

The continuous random variable X has PDF $\mathrm{f}(x)$ given by

$$\mathrm{f}(x) = \begin{cases} \dfrac{x}{50} & \text{for } 0 \leqslant x \leqslant 10 \\ 0 & \text{otherwise.} \end{cases}$$

(i) Find $\mathrm{E}(3X + 4)$.

(ii) Find $3\mathrm{E}(X) + 4$.

(iii) Find $\mathrm{Var}(3X + 4)$.

(iv) Verify that $\mathrm{Var}(3X + 4) = 3^2\mathrm{Var}(X)$.

→

Solution

(i) $\quad E(3X + 4) = \int_0^{10} (3x + 4)\dfrac{x}{50}\,dx \longleftarrow$

Here you are using
$$E(g(X)) = \int_{\substack{\text{All} \\ \text{value} \\ \text{of } x}} g(x)f(x)\,dx$$

$$= \int_0^{10} \dfrac{1}{50}(3x^2 + 4x)\,dx$$

$$= \left[\dfrac{x^3}{50} + \dfrac{x^2}{25}\right]_0^{10}$$

$$= 20 + 4$$

$$= 24$$

(ii) $\quad 3E(X) + 4 = 3\int_0^{10} x\dfrac{x}{50}\,dx + 4$

Here you are using
$$\text{Var}(g(X)) = \int (g(x))^2 f(x)\,dx$$
$$- \{E(g(X))\}^2$$
and from part (i) you know that
$E(3X + 4) = 24$

$$= \left[\dfrac{3}{150}x^3\right]_0^{10} + 4$$

$$= 20 + 4$$

$$= 24$$

Notice here that $E(3X + 4) = 24 = 3E(X) + 4$.

(iii) To find $\text{Var}(3X + 4)$, use

$$\text{Var}(3X + 4) = \int_0^{10} (3x + 4)^2 \dfrac{1}{50}x\,dx - 24^2 \longleftarrow$$

You then multiply out the brackets and then multiply by x

$$= \int_0^{10} \dfrac{1}{50}(9x^3 + 24x^2 + 16x)\,dx - 576$$

$$= \dfrac{1}{50}\left[\dfrac{9x^4}{4} + 8x^3 + 8x^2\right]_0^{10} - 576$$

$$= \dfrac{1}{50}\left[\dfrac{9 \times 10^4}{4} + 8 \times 10^3 + 8 \times 10^2\right] - 576$$

$$= \dfrac{1}{50} \times 31300 - 576$$

$$= 50$$

You may recall from A Level Mathematics
$$E(X) = \int_{\substack{\text{All} \\ \text{values} \\ \text{of } x}} x\, f(x)\,dx$$

$$E(X^2) = \int_{\substack{\text{All} \\ \text{values} \\ \text{of } x}} x^2\, f(x)\,dx$$

(iv) $\quad \text{Var}(X) = E(X^2) - [E(X)]^2$

$$E(X^2) = \int_0^{10} x^2 \dfrac{1}{50}x\,dx \qquad E(X) = \int_0^{10} x\dfrac{1}{50}x\,dx$$

$$E(X^2) = \int_0^{10} \dfrac{1}{50}x^3\,dx \qquad E(X) = \int_0^{10} \dfrac{1}{50}x^2\,dx$$

$$E(X^2) = \left[\frac{1}{200}x^4\right]_0^{10} \qquad E(X) = \left[\frac{1}{150}x^3\right]_0^{10}$$

$$E(X^2) = 50 \qquad\qquad E(X) = 6.\dot{6}$$

$$\text{Var}(X) = 50 - 6.\dot{6}^2 = 5.\dot{5} \longleftarrow$$

Here you are using

$$\text{Var}(X) = E(X^2) - [E(X)]^2$$

$$3^2\text{Var}(X) = 9 \times 5.\dot{5} = 50$$

From part (iii), Var$(3X + 4) = 50$

So Var$(3X + 4) = 3^2$ Var(X) as required.

Exercise 1A

1 A random variable X has PDF

$$f(x) = \begin{cases} kx & \text{for } 0 \leq x < 5 \\ k(10 - x) & \text{for } 5 \leq x \leq 10 \\ 0 & \text{otherwise.} \end{cases}$$

(i) Find the value of the constant k.

(ii) Sketch $y = f(x)$.

(iii) Find $P(4 \leq X \leq 6)$.

2 A random variable X has PDF

$$f(x) = \begin{cases} (x - 1)(2 - x) & \text{for } 1 \leq x < 2 \\ a & \text{for } 2 \leq x \leq 4 \\ 0 & \text{otherwise.} \end{cases}$$

(i) Find the value of the constant a.

(ii) Sketch $y = f(x)$.

(iii) Find $P(1.5 \leq X \leq 2.5)$.

(iv) Find $P(|X - 2| < 1)$.

3 The random variable X has the following PDF

$$f(x) = \begin{cases} 0.01(a + x) & \text{for } -a \leq x < 0 \\ 0.01(a - x) & \text{for } 0 \leq x \leq a \\ 0 & \text{otherwise.} \end{cases}$$

(i) Find the value of a.

(ii) Find $P(3 \leq X \leq 5)$.

(iii) Find $P(|X| < 4)$.

4   A continuous random variable X has the PDF

$$f(x) = \begin{cases} k & \text{for } 0 \leqslant x \leqslant 5 \\ 0 & \text{otherwise.} \end{cases}$$

(i)   Find the value of k.

(ii) Sketch the graph of f(x).

(iii) Find E(X).

(iv) Find E($4X - 3$) and show that your answer is the same as $4E(X) - 3$.

5   The continuous random variable X has PDF

$$f(x) = \begin{cases} 4x^3 & \text{for } 0 \leqslant x \leqslant 1 \\ 0 & \text{otherwise.} \end{cases}$$

(i)   Find E(X).

(ii) Find E(X^2).

(iii) Find Var(X).

(iv) Verify that $E(5X + 1) = 5E(X) + 1$.

6   The number of kilograms of metal extracted from 10 kg of ore from a certain mine is modelled by a continuous random variable X with probability density function f(x), where f(x) = $cx(2 - x)^2$ if $0 \leqslant x \leqslant 2$ and f(x) = 0 otherwise, where c is a constant.

Show that c is $\dfrac{3}{4}$, and find the mean and variance of X.

The cost of extracting the metal from 10 kg of ore is $10x$. Find the expected cost of extracting the metal from 10 kg of ore.

7   A continuous random variable Y has PDF

$$f(y) = \begin{cases} \dfrac{2}{9} y(3 - y) & \text{for } 0 \leqslant y \leqslant 3 \\ 0 & \text{otherwise.} \end{cases}$$

(i)   Find E(Y).

(ii) Find E(Y^2).

(iii) Find $E(Y^2) - (E(Y))^2$.

(iv) Find $E(2Y^2 + 3Y + 4)$.

(v)   Find $\displaystyle\int_0^3 (y - E(Y))^2 \, f(y) \, dy$.

Why is the answer the same as that for part (iii)?

8 A continuous random variable X has PDF f(x), where

$$f(x) = \begin{cases} 12x^2(1-x) & \text{for } 0 \leqslant x \leqslant 1 \\ 0 & \text{otherwise.} \end{cases}$$

(i) Find μ, the mean of X.

(ii) Find $\text{E}(6X-7)$ and show that your answer is the same as $6\text{E}(X) - 7$.

(iii) Find the standard deviation of X.

(iv) What is the probability that a randomly selected value of X lies within one standard deviation of μ?

9 The continuous random variable X has PDF

$$f(x) = \begin{cases} \dfrac{2}{25}(7-x) & \text{for } 2 \leqslant x \leqslant 7 \\ 0 & \text{otherwise.} \end{cases}$$

The function g(x) is defined by g$(x) = 3x^2 + 4x + 7$.

(i) Find $\text{E}(X)$.

(ii) Find $\text{E}(\text{g}(X))$.

(iii) Find $\text{E}(X^2)$ and hence find $3\text{E}(X^2) + 4\text{E}(X) + 7$.

(iv) Use your answers to parts (ii) and (iii) to verify that $\text{E}(\text{g}(X)) = 3\text{E}(X^2) + 4\text{E}(X) + 7$.

10 A toy company sells packets of coloured plastic equilateral triangles. The triangles are actually offcuts from the manufacture of a totally different toy, and the length, X, of one side of a triangle may be modelled as a random variable with a uniform (rectangular) distribution for $2 \leqslant x \leqslant 8$.

(i) Find the PDF of X.

(ii) An equilateral triangle of side x has area a. Find the relationship between a and x.

(iii) Find the probability that a randomly selected triangle has area greater than $15\,\text{cm}^2$.

   Tip: What does this imply about x?

(iv) Find the expectation and variance of the area of a triangle.

Answers to exercises are available at www.hoddereducation.com/cambridgeextras

1.3 The cumulative distribution function

500 enter community half-marathon

There was a record entry for this year's community half-marathon, including several famous names who were treating it as a training run in preparation for the Cape Town marathon. Overall winner was Reuben Mhango in 1 hour 4 minutes and 2 seconds; the first woman home was 37-year-old Kagendo Govender in 1 hour 20 minutes exactly. There were many fun runners but everybody completed the course within 4 hours.

$150 prize to be won

Marius Erasmus, chair of the Half Committee, says: 'This year we restricted entries to 500 but this meant disappointing many people. Next year we intend to allow everybody to run and expect a much bigger entry. In order to allow us to marshal the event properly we need a statistical model to predict the flow of runners, and particularly their finishing times. We are offering a prize of $150 for the best such model submitted.

Time (hours)	Finished (%)
$1\frac{1}{4}$	3
$1\frac{1}{2}$	15
$1\frac{3}{4}$	33
2	49
$2\frac{1}{4}$	57
$2\frac{1}{2}$	75
3	91
$3\frac{1}{2}$	99
4	100

An entrant for the competition proposes a model in which a runner's time, X hours, is a continuous random variable with PDF

$$f(x) = \begin{cases} \dfrac{4}{27}(x-1)(4-x)^2 & \text{for } 1 \leqslant x \leqslant 4 \\ 0 & \text{otherwise.} \end{cases}$$

According to this model, the mode is at 2 hours, and everybody's finishing times were between 1 hour and 4 hours; see Figure 1.3.

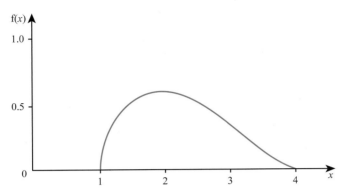

▲ **Figure 1.3**

How does this model compare with the figures you were given for the actual race?

Those figures gave the **cumulative distribution**, the total numbers (expressed as percentages) of runners who had finished by certain times. To obtain the equivalent figures from the model, you want to find an expression for $P(X \leqslant x)$. The function giving $P(X \leqslant x)$ is called the **cumulative distribution function (CDF)**, and it is usually denoted by $F(x)$. The best method of obtaining the cumulative distribution function is to use indefinite integration.

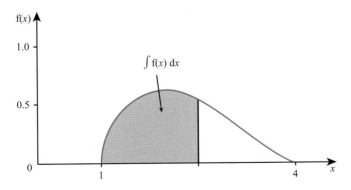

▲ **Figure 1.4**

Answers to exercises are available at www.hoddereducation.com/cambridgeextras

You want to find the area under $y = f(x)$ up to a particular value of x. So you write

$$F(x) = P(X \leqslant x)$$

$$= \int f(x)\,dx$$

$$= \int \frac{4}{27}(x-1)(4-x)^2\,dx$$

$$= \frac{4}{27}\int (x^3 - 9x^2 + 24x - 16)\,dx$$

$$= \frac{4}{27}\left(\frac{1}{4}x^4 - 3x^3 + 12x^2 - 16x\right) + c$$

$$= \frac{1}{27}x^4 - \frac{4}{9}x^3 + \frac{16}{9}x^2 - \frac{64}{27}x + c$$

But $P(X \leqslant 1) = 0$, so $F(1) = 0$ and you can use this to find the value of c:

No runners finish in less than 1 hour, so if x is less than 1, the probability of a runner finishing in less than x hours is 0.

$$\frac{1}{27} - \frac{4}{9} + \frac{16}{9} - \frac{64}{27} + c = 0$$

$$\Rightarrow \qquad\qquad c = 1$$

Hence the cumulative distribution function is given by

All runners finish in less than 4 hours, so if x is greater than 4, the probability of a runner finishing in less than x hours is 1.

$$F(x) = \begin{cases} 0 & \text{for } x < 1 \\ \dfrac{1}{27}x^4 - \dfrac{4}{9}x^3 + \dfrac{16}{9}x^2 - \dfrac{64}{27}x + 1 & \text{for } 1 \leqslant x \leqslant 4 \\ 1 & \text{for } x > 4. \end{cases}$$

It is possible to use definite integration, but this causes a problem as you cannot use the same letter for both a limit of the integral and as the variable of integration. So you would have to change the variable of integration, which is a **dummy variable** as it does not appear in the final answer, to a different letter.

To find the proportions of runners finishing by any time, substitute that value for x; so, when $x = 2$

You would not be correct to write down an expression like $$F(x) = \int_1^x \frac{4}{27}(x-1)(4-x)^2\,dx$$ INCORRECT since x would then be both a limit of the integral and the variable used within it. To overcome this problem you use a dummy variable, say, u, so that $F(x)$ is now written $$F(x) = \int_1^x \frac{4}{27}(u-1)(4-u)^2\,dt$$ CORRECT

$$F(2) = \frac{1}{27} \times 2^4 - \frac{4}{9} \times 2^3 + \frac{16}{9} \times 2^2 - \frac{64}{27} \times 2 + 1$$

$$= 0.41 \text{ to two decimal places.}$$

Here is the complete table, with all the values worked out.

Time (hours)	Model	Runners
1.00	0.00	0.00
1.25	0.04	0.03
1.50	0.13	0.15
1.75	0.26	0.33
2.00	0.41	0.49
2.25	0.55	0.57
2.50	0.69	0.75
3.00	0.89	0.91
3.50	0.98	0.99
4.00	1.00	1.00

Notice the distinctive shape of the curves of these functions (Figure 1.5), sometimes called an **ogive**.

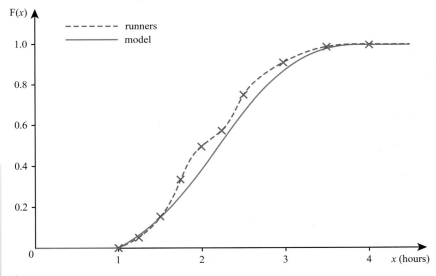

▲ **Figure 1.5**

> **Note**
>
> You have probably met this shape already when drawing cumulative frequency curves.

> **?**
>
> › Do you think that this model is worth the $150 prize? If you were on the organising committee what more might you look for in a model?

Properties of the cumulative distribution function, $F(x)$

The graphs on the next page, Figure 1.6, show the probability density function $f(x)$ and the cumulative distribution function $F(x)$ of a typical continuous random variable X.

You will see that the values of the random variable always lie between a and b.

Answers to exercises are available at www.hoddereducation.com/cambridgeextras

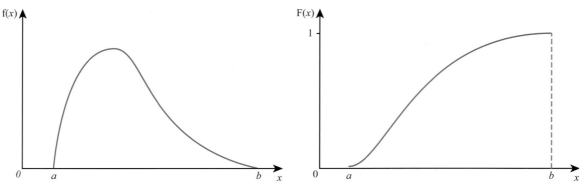

▲ **Figure 1.6**

These graphs illustrate a number of general results for cumulative distribution functions.

1 $F(x) = 0$ for $x \leqslant a$, the lower limit of x.

   The probability of X taking a value less than or equal to a is zero; the value of X must be greater than or equal to a.

2 $F(x) = 1$ for $x \geqslant b$, the upper limit of x. X cannot take values greater than b.

3 $P(c \leqslant X \leqslant d) = F(d) - F(c)$

   $P(c \leqslant X \leqslant d) = P(X \leqslant d) - P(X \leqslant c)$

   This is very useful when finding probabilities from a PDF or a CDF.

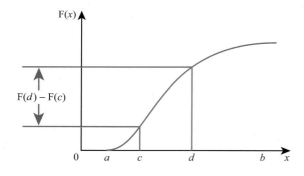

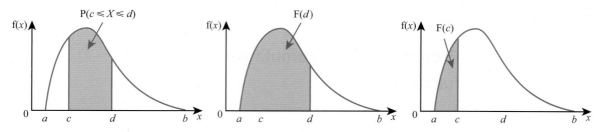

▲ **Figure 1.7**

4 The median, m, satisfies the equation $F(m) = 0.5$.

$P(X \leqslant m) = 0.5$ by definition of the median.

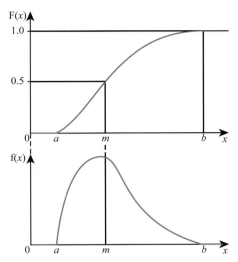

▲ **Figure 1.8**

5 $f(x) = \dfrac{d}{dx} F(x) = F'(x)$

Since you integrate $f(x)$ to obtain $F(x)$, the reverse must also be true: differentiating $F(x)$ gives $f(x)$.

6 $F(x)$ is a continuous function: the graph of $y = F(x)$ has no gaps.

> ### Notes
>
> **1** Notice the use of lower and upper case letters here. The probability density function is denoted by the lower case f, whereas the cumulative distribution function is given the upper case F.
>
> **2** The cumulative distribution function can be found by integrating the PDF f(x). So $F(x) = P(X \leqslant x) = \displaystyle\int_{-\infty}^{x} f(t)\, dt$
>
> **3** The term 'cumulative distribution function' is often abbreviated to CDF.

Example 1.4

A machine saws planks of wood to a nominal length. The continuous random variable X represents the error in millimetres of the actual length of a plank coming off the machine. The variable X has PDF $f(x)$ where

$$f(x) = \begin{cases} \dfrac{10 - x}{50} & \text{for } 0 \leqslant x \leqslant 10 \\ 0 & \text{otherwise} \end{cases}$$

(i)   Sketch $f(x)$.

(ii) Find the cumulative distribution function $F(x)$.

(iii) Sketch $F(x)$ for $0 \leqslant x \leqslant 10$.

Answers to exercises are available at www.hoddereducation.com/cambridgeextras

(iv)   Find $P(2 \leq X \leq 7)$.

(v)   Find the median value of X.

A customer refuses to accept planks for which the error is greater than 8 mm.

(vi)   What percentage of planks will he reject?

Solution

(i)

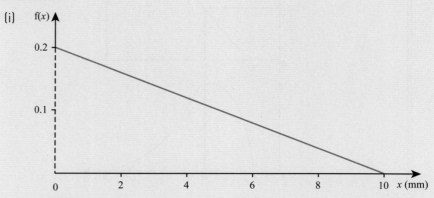

▲ Figure 1.9

(ii) $F(x) = \displaystyle\int_0^x \frac{(10 - u)}{50}\,du$

$\quad = \dfrac{1}{50}\left[10u - \dfrac{u^2}{2}\right]_0^x$

$\quad = \dfrac{1}{5}x - \dfrac{1}{100}x^2$

The full definition of $F(x)$ is:

$$F(x) = \begin{cases} 0 & \text{for } x < 0 \\ \dfrac{1}{5}x - \dfrac{1}{100}x^2 & \text{for } 0 \leq x \leq 10 \\ 1 & \text{for } x > 10. \end{cases}$$

(iii)   The graph $F(x)$ is shown in Figure 1.10.

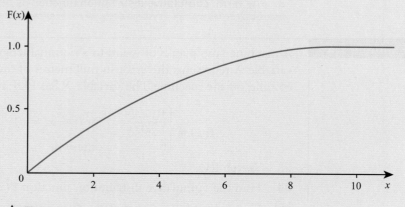

▲ Figure 1.10

(iv)

$$P(2 \leqslant X \leqslant 7) = F(7) - F(2)$$

$$= \left[\frac{7}{5} - \frac{49}{100}\right] - \left[\frac{2}{5} - \frac{4}{100}\right]$$

$$= 0.91 - 0.36$$

$$= 0.55$$

(v) The median value of X is found by solving the equation

$$F(m) = 0.5$$

$$\frac{1}{5}m - \frac{1}{100}m^2 = 0.5.$$

This is rearranged to give

$$m^2 - 20m + 50 = 0$$

$$m = \frac{20 \pm \sqrt{20^2 - 4 \times 50}}{2}$$

$$m = 2.93 \quad \text{(or 17.07, outside the domain for } X\text{)}.$$

The median error is 2.93 mm.

(vi) The customer rejects those planks for which $8 \leqslant X \leqslant 10$.

$$P(8 \leqslant X \leqslant 10) = F(10) - F(8)$$

$$= 1 - 0.96$$

so 4% of planks are rejected.

Example 1.5

The PDF of a continuous random variable X is given by

$$f(x) = \begin{cases} \dfrac{x}{24} & \text{for } 0 \leqslant x \leqslant 4 \\[2mm] \dfrac{(12-x)}{48} & \text{for } 4 \leqslant x \leqslant 12 \\[2mm] 0 & \text{otherwise.} \end{cases}$$

(i) Sketch $f(x)$.

(ii) Find the cumulative distribution function $F(x)$.

(iii) Sketch $F(x)$.

Solution

(i) The graph of $f(x)$ is shown in Figure 1.11.

→

Answers to exercises are available at www.hoddereducation.com/cambridgeextras

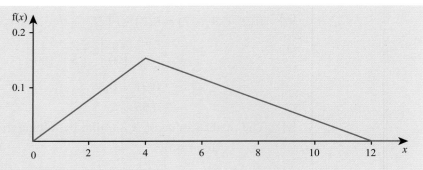

▲ **Figure 1.11**

(ii)   For $0 \leqslant x \leqslant 4$, $F(x) = \int_0^x \frac{u}{24} \, du$

$$= \left[\frac{u^2}{48} \right]_0^x$$

$$= \frac{x^2}{48}$$

and so $F(4) = \frac{1}{3}$.

For $4 \leqslant x \leqslant 12$, a second integration is required:

$$F(x) = \int_0^4 \frac{u}{24} \, du + \int_4^x \left(\frac{12 - u}{48} \right) du$$

$$= F(4) + \left[\frac{u}{4} - \frac{u^2}{96} \right]_4^x$$

$$= \frac{1}{3} + \frac{x}{4} - \frac{x^2}{96} - \frac{5}{6}$$

$$= -\frac{1}{2} + \frac{x}{4} - \frac{x^2}{96}.$$

So the full definition of $F(x)$ is

$$F(x) = \begin{cases} 0 & \text{for } x < 0 \\ \dfrac{x^2}{48} & \text{for } 0 \leqslant x \leqslant 4 \\ -\dfrac{1}{2} + \dfrac{x}{4} - \dfrac{x^2}{96} & \text{for } 4 \leqslant x \leqslant 12 \\ 1 & \text{for } x > 12. \end{cases}$$

(iii) The graph of F(x) is shown in Figure 1.12.

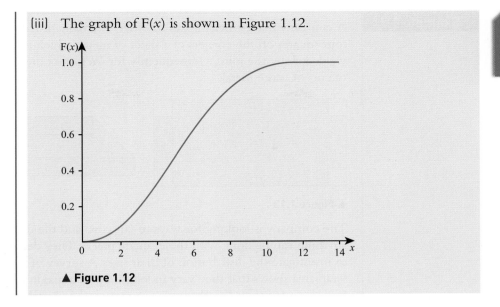

▲ **Figure 1.12**

| Example 1.6 | The continuous random variable X has cumulative distribution function $F(x)$ given by |

$$F(x) = \begin{cases} 0 & \text{for } x < 2 \\ \dfrac{x^2}{32} - \dfrac{1}{8} & \text{for } 2 \leqslant x \leqslant 6 \\ 1 & \text{for } x > 6. \end{cases}$$

Find the PDF $f(x)$.

Solution

$$f(x) = \frac{\mathrm{d}}{\mathrm{d}x} F(x)$$

$$f(x) = \begin{cases} \dfrac{\mathrm{d}}{\mathrm{d}x} F(x) = 0 & \text{for } x < 2 \\ \dfrac{\mathrm{d}}{\mathrm{d}x} F(x) = \dfrac{x}{16} & \text{for } 2 \leqslant x \leqslant 6 \\ \dfrac{\mathrm{d}}{\mathrm{d}x} F(x) = 0 & \text{for } x > 6. \end{cases}$$

1.4 Finding the PDF of a function of a continuous random variable

The cumulative distribution function provides you with a stepping stone between the PDF of a continuous random variable and that of a function of that variable. Example 1.7 shows how it is done.

Answers to exercises are available at www.hoddereducation.com/cambridgeextras

Example 1.7

A company makes metal boxes to order. The basic process consists of cutting four squares off the corners of a sheet of metal, which is then folded and welded along the joins. Consequently, for every box there are four square offcuts of waste metal.

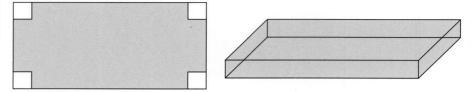

▲ **Figure 1.13**

The company is looking for ways to cut costs and the designers wonder if anything can be done with these square pieces. They decide in the first place to investigate the distribution of their sizes. A survey of the large pile in their scrap area shows that they vary in length up to a maximum of 2 decimetres. It is suggested their lengths in decimetres can be modelled as a continuous random variable L with probability density function

$$f(l) = \begin{cases} \dfrac{l}{4}(4 - l^2) & \text{for } 0 \leqslant l \leqslant 2 \\ 0 & \text{otherwise.} \end{cases}$$

Assume this model to be accurate.

(i)   Find the cumulative distribution function for the length of a square.

(ii) Hence derive the cumulative distribution function for the area of a square.

(iii) Find the PDF for the area of a square.

(iv) Sketch the graphs of the probability density functions and the cumulative distribution functions of the length and the area.

(v)   Find the mean area of the square offcuts when making a box.

Solution

(i)   The CDF is

$$F(l) = \int_0^l \frac{u}{4}(4 - u^2)\,du$$

Notice the use of u as a dummy variable for L.

$$= \left[\frac{u^2}{2} - \frac{u^4}{16} \right]_0^l$$

and so

$$F(l) = \begin{cases} 0 & \text{for } l \leqslant 0 \\ \dfrac{l^2}{2} - \dfrac{l^4}{16} & \text{for } 0 < l \leqslant 2 \\ 1 & \text{for } l < 2. \end{cases}$$

(ii) The area, a, and length, l, of a square are related by

$$a = l^2$$

and since $0 < l \leqslant 2$

it follows that $0 < a \leqslant 2^2$; that is, $0 < a \leqslant 4$.

Substituting a for l^2 in the answer to part (i), and using the appropriate range of values for a, gives the cumulative distribution function $H(a)$, because

$$\begin{aligned} H(a) &= P(A \leqslant a) \\ &= P(L^2 \leqslant l^2) \\ &= P(L \leqslant l) \qquad \text{(as } L > 0\text{)} \\ &= F(l). \end{aligned}$$

Since $\qquad F(l) = \dfrac{l^2}{2} - \dfrac{l^4}{16} \qquad \text{for } 0 < l \leqslant 2$

it follows that $\qquad H(a) = \begin{cases} \dfrac{a}{2} - \dfrac{a^2}{16} & \text{for } 0 < a \leqslant 4 \\ 1 & \text{for } a > 4. \end{cases}$

(iii) The PDF for the area of a square is found by differentiating $H(a)$

$$h(a) = \begin{cases} \dfrac{d}{da} H(a) = \dfrac{1}{2} - \dfrac{a}{8} & \text{for } 0 < a \leqslant 4 \\ 0 & \text{otherwise.} \end{cases}$$

> ## Note
> Notice the use of H and h for the CDF and the PDF of the area, in place of F and f. The different letters are used to distinguish these from the corresponding functions for the length.

(iv) The graphs of the PDFs of the length and the area are shown in Figure 1.14.

(a)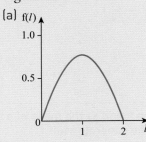

(b)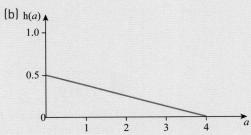

▲ **Figure 1.14**

(a) $f(l) = \begin{cases} \dfrac{1}{4} l(4 - l^2) & \text{for } 0 < l \leqslant 2 \\ 0 & \text{otherwise.} \end{cases}$

Answers to exercises are available at www.hoddereducation.com/cambridgeextras

(b) $h(a) = \begin{cases} \dfrac{1}{2} - \dfrac{a}{8} & \text{for } 0 < a \leqslant 4 \\ 0 & \text{otherwise.} \end{cases}$

The graphs of the CDFs of the length and the area are shown in Figure 1.15.

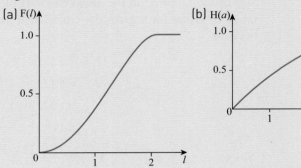

(a) F(l)

(b) H(a)

▲ Figure 1.15

(a) $F(l) = \begin{cases} \dfrac{l^2}{2} - \dfrac{l^4}{16} & \text{for } 0 < l \leqslant 2 \\ 1 & \text{for } l > 2. \end{cases}$

(b) $H(a) = \begin{cases} 0 & \text{for } a \leqslant 0 \\ \dfrac{a}{2} - \dfrac{a^2}{16} & \text{for } 0 \leqslant a \leqslant 4 \\ 1 & \text{for } a > 4. \end{cases}$

(v) $\begin{aligned} \text{Mean} = E(A) &= \int_0^4 a h(a)\, da \\ &= \int_0^4 \left(\frac{a}{2} - \frac{a^2}{8} \right) da \\ &= \left[\frac{a^2}{4} - \frac{a^3}{24} \right]_0^4 \\ &= \frac{4}{3} \end{aligned}$

Note

This could also have been found as the mean of a function of a continuous random variable, using the general result

$$E[g(X)] = \int_{\substack{\text{All} \\ \text{values} \\ \text{of } x}} g(x) f(x)\, dx$$

where x is the length (not the area) of one of the squares.

In this case, $g(x) = x^2$ $f(x) = \frac{x}{4}(4 - x^2)$ and $0 < X \leq 2$

giving
$$E[g(X)] = \int_0^2 x^2 \frac{x}{4}(4 - x^2)\,dx$$

$$= \int_0^2 \left(x^3 - \frac{x^5}{4}\right)dx$$

$$= \left[\frac{x^4}{4} - \frac{x^6}{24}\right]_0^2$$

$$= 4 - \frac{64}{24} = \frac{4}{3}$$

i.e. the same answer.

Exercise 1B

1   The continuous random variable X has PDF $f(x)$ where

$$f(x) = \begin{cases} 0.2 & \text{for } 0 \leq x \leq 5 \\ 0 & \text{otherwise.} \end{cases}$$

(i)   Find $E(X)$.

(ii) Find the cumulative distribution function, $F(x)$.

(iii) Find $P(0 \leq x \leq 2)$ using

   (a) $F(x)$

   (b) $f(x)$

   and show your answer is the same by each method.

2   The continuous random variable U has PDF $f(u)$ where

$$f(u) = \begin{cases} ku & \text{for } 5 \leq u \leq 8 \\ 0 & \text{otherwise.} \end{cases}$$

(i)   Find the value of k.

(ii) Sketch $f(u)$.

(iii) Find $F(u)$.

(iv) Sketch the graph of $F(u)$.

3   A continuous random variable X has PDF $f(x)$ where

$$f(x) = \begin{cases} cx^2 & \text{for } 1 \leq x \leq 4 \\ 0 & \text{otherwise.} \end{cases}$$

(i)   Find the value of c.

(ii) Find $F(x)$.

(iii) Find the median of X.

(iv) Find the mode of X.

Answers to exercises are available at <u>www.hoddereducation.com/cambridgeextras</u>

4 The continuous random variable X has PDF $f(x)$ given by

$$f(x) = \begin{cases} \dfrac{k}{(x+1)^4} & \text{for } x \geqslant 0 \\ 0 & \text{otherwise,} \end{cases}$$

where k is a constant.

(i) Show that $k = 3$, and find the cumulative distribution function.

(ii) Find also the value of x such that $P(X < x) = \dfrac{7}{8}$.

5 The continuous random variable X has CDF given by

$$F(x) = \begin{cases} 0 & \text{for } x < 0 \\ 2x - x^2 & \text{for } 0 \leqslant x \leqslant 1 \\ 1 & \text{for } x > 1. \end{cases}$$

(i) Find $P(X > 0.5)$.

(ii) Find the value of q such that $P(X < q) = \dfrac{1}{4}$.

(iii) Find the PDF $f(x)$ of X, and sketch its graph.

6 The continuous random variable X has PDF $f(x)$ given by

$$f(x) = \begin{cases} k(9 - x^2) & \text{for } 0 \leqslant x \leqslant 3 \\ 0 & \text{otherwise,} \end{cases}$$

where k is a constant.

Show that $k = \dfrac{1}{18}$ and find the values of $E(X)$ and $Var(X)$.

Find the cumulative distribution function for X, and verify by calculation that the median value of X is between 1.04 and 1.05.

7 A continuous random variable X has PDF $f(x)$ where

$$f(x) = \begin{cases} 6x(1 - x) & \text{for } 0 \leqslant x \leqslant 1 \\ 0 & \text{otherwise.} \end{cases}$$

Find μ, the mean of X, and show that σ^2, the variance of X, is 0.05.

Show that $F(x)$, the probability that $X \leqslant x$ (for any value of x between 0 and 1), satisfies

$$F(x) = \begin{cases} 0 & \text{for } x < 0 \\ 3x^2 - 2x^3 & \text{for } 0 \leqslant x \leqslant 1 \\ 1 & \text{for } x > 1. \end{cases}$$

Use this result to show that $P(|X - \mu| < \sigma^2) = 0.1495$.

What would this probability be if, instead, X were normally distributed?

8 The continuous random variable X has PDF $f(x)$ given by

$$f(x) = \begin{cases} 4x^3 & \text{for } 0 \leqslant x \leqslant 1 \\ 0 & \text{otherwise.} \end{cases}$$

(i)   Find F(x).

The continuous random variable Y is given by $Y = X^3$. The cumulative distribution function of Y is denoted by H(y).

(ii) Find H(y).

(iii) Find h(y).

(iv) Find P($X < 0.5$).

(v)   Find P($Y < 0.5$).

9   The continuous random variable X has PDF f(x) given by

$$f(x) = \begin{cases} 2(1-x) & \text{for } 0 \leqslant x \leqslant 1 \\ 0 & \text{otherwise.} \end{cases}$$

The continuous random variable Y is given by $Y = (1 - X)^2$.

(i)   Find h(y), the probability density function of Y and name the distribution that it represents.

(ii) Find P($X < 0.9$).

(iii) Find P($Y < 0.9$).

PS   10 The continuous random variable X has PDF f(x) given by

$$f(x) = \begin{cases} 0.1 & \text{for } 0 \leqslant x \leqslant 10 \\ 0 & \text{otherwise.} \end{cases}$$

The continuous random variable Y is given by $Y = X^4$.

(i)   Find H(y) where H(y) is the cumulative distribution function of Y.

(ii)   Find h(y).

(iii) Find P($X < 5$).

(iv)   Find P($Y < 5000$).

The continuous random variable Z is given by $Z = X^p$ where $p > 0$.

(v)   Find J(z) where J(z) is the cumulative distribution function of Z.

(vi)   Find j(z).

(vii) Find P($Z < 5^p$).

CP   11 The continuous random variable X has PDF f(x) given by

$$f(x) = \begin{cases} 0.2 & \text{for } 3 \leqslant x \leqslant 8 \\ 0 & \text{otherwise.} \end{cases}$$

The continuous random variable Y is given by $Y = 5X^2$.

(i)   Find H(y) where H(y) is the cumulative distribution function of Y.

(ii)   Find h(y).

(iii) Find P($Y < 100$).

(iv)   State the median value of X.

(v)   Use H(y) to find the median value of Y, showing your working.

(vi)   Show that you can use the equation $Y = 5X^2$ to find the median value of Y from that of X.

(vii) Write down the value of E(X) and find E(Y). Can you use the equation $Y = 5X^2$ to find the mean value of Y from that of X?

Answers to exercises are available at www.hoddereducation.com/cambridgeextras

KEY POINTS

1   If X is a continuous random variable with probability density function (PDF) f(x) then

- $$\int f(x)\,dx = 1$$

- $f(x) \geqslant 0$ for all x

- $$P(c \leqslant X \leqslant d) = \int_c^d f(x)\,dx$$

- $$E(X) = \int x f(x)\,dx$$

- $$\mathrm{Var}(X) = \int x^2 f(x)\,dx - \left[E(X)\right]^2$$

- the median m of X is the value for which
  $$\int_{-\infty}^m f(x)\,dx \text{ and } \int_m^\infty f(x)\,dx = 0.5$$

- the mode of X is the value for which f(x) has its greatest magnitude.

A probability density function may be defined piecewise.

2   If g$[X]$ is a function of X then the expectation and variance of X are

- $$E\big(g(X)\big) = \int g(x)f(x)\,dx$$

- $$\mathrm{Var}\big(g(X)\big) = \int (g(x))^2\, f(x)\,dx - \big(E\big(g(X)\big)\big)^2.$$

3   If f(x) is the probability density function of X then the cumulative distribution function (CDF) of X is F(x) where

- $$F(x) = \int_a^x f(u)\,du \text{ where the constant } a \text{ is the lower limit of } X$$

- $$f(x) = \frac{d}{dx}F(x)$$

- for the median, m, F$(m) = 0.5$.

4   Given that f(x) is the probability density function of X, you can find that of a related variable (e.g. $Y = X^2$).

To do this you need first to find the cumulative distribution function of X and use this to find that of Y. You can then differentiate the cumulative distribution function of Y to find f(y), the probability density function of Y.

LEARNING OUTCOMES

Now that you have finished this chapter, you should be able to

- use a probability density function that may be defined piecewise

- find the mean and variance from a given PDF

- use the general result $E\big(g(X)\big) = \int g(x)f(x)\,dx$ where f(x) is the probability density function of the continuous random variable X, and g(X) is a function of X

- use the general result $Var\big(g(X)\big) = \int \big(g(x)\big)^2 f(x)\,dx - \big\{E\big(g(X)\big)\big\}^2$ where f(x) is the probability density function of the continuous random variable X, and g(X) is a function of X

- understand and use the relationship between the probability density function (PDF) and the cumulative distribution function (CDF)

- use a PDF or a CDF to evaluate probabilities

- use a PDF or a CDF to calculate the median and other percentiles

- use cumulative distribution functions of related variables.

Answers to exercises are available at <u>www.hoddereducation.com/cambridgeextras</u>

2 Inference using normal and *t*-distributions

2.1 Interpreting sample data using the *t*-distribution

Students find new bat

Two students and a lecturer have found their way into the textbooks. On a recent field trip they discovered a small colony of a previously unknown bat living in a cave.

'Somewhere in Northern India' is all that Shakila Mahadavan, 20, would say about its location. 'We don't want the general public disturbing the bats or worse still catching them for specimens,' she explained.

The other two members of the group, lecturer Jaswinder Pal and 21-year-old Vijay Kumar, showed lots of photographs of the bats as well as pages of measurements that they had gently made on the few bats they had caught before releasing them back into their cave.

The measurements referred to in the article include the weights (in g) of eight bats that were identified as adult males.

| 156 | 132 | 160 | 142 | 145 | 138 | 151 | 144 |

From these figures, the team want to estimate the mean weight of an adult male bat, and 95% confidence limits for their figure.

It is clear from the newspaper report that these are the only measurements available. All that is known about the parent population is what can be inferred from these eight measurements. You know neither the mean nor the standard deviation of the parent population, but you can estimate both.

The mean is estimated to be the same as the sample mean

$$\frac{156 + 132 + 160 + 142 + 145 + 138 + 151 + 144}{8} = 146.$$

When it comes to estimating the standard deviation, start by finding the sample variance

$$s^2 = \frac{S_{xx}}{n-1} = \sum_i \frac{(x_i - \bar{x})^2}{(n-1)}$$

and then take the square root to find the standard deviation, s.

The use of $(n-1)$ as divisor illustrates the important concept of degrees of freedom.

The deviations of the eight numbers are as follows.

$156 - 146 = 10$

$132 - 146 = -14$

$160 - 146 = 14$

$142 - 146 = -4$

$145 - 146 = -1$

$138 - 146 = -8$

$151 - 146 = 5$

$144 - 146 = -2$

Note

The deviation is the difference (+ or −) of the value from the mean. In this example the mean is 146.

Note

You need to know the degrees of freedom in many situations where you are calculating confidence intervals or conducting hypothesis tests.

These eight deviations are not independent: they must add up to zero because of the way the mean is calculated. This means that when you have worked out the first seven deviations, it is inevitable that the final one has the value it does (in this case −2). Only seven values of the deviation are independent and, in general, only $(n-1)$ out of the n deviations from the sample mean are independent.

Consequently, there are $n-1$ **free variables** in this situation. The number of free variables within a system is called the **degrees of freedom** and denoted by v.

So the sample variance is worked out using divisor $(n-1)$. The resulting value is very useful because it is an **unbiased estimate of the parent population variance**.

In the case of the bats, the estimated population variance is

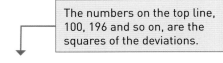
The numbers on the top line, 100, 196 and so on, are the squares of the deviations.

Note

A particular value of the sample variance is denoted by s^2, the associated random variable by S^2.

$$s^2 = \frac{(100 + 196 + 196 + 16 + 1 + 64 + 25 + 4)}{7} = 86$$

and the corresponding value of the standard deviation is $s = \sqrt{86} = 9.27$.

Calculating the confidence intervals

Returning to the problem of estimating the mean weight of the bats, you now know that

$$\bar{x} = 146, \quad s^2 = 86, \quad s = 9.27 \quad \text{and} \quad \nu = 8 - 1 = 7$$

Before starting on further calculations, there are some important and related points to notice.

1 This is a small sample. It would have been much better if they had managed to catch and weigh more than eight bats.

2 The true parent standard deviation, σ, is unknown and, consequently, the standard deviation of the sampling distribution given by the central limit theorem, $\frac{\sigma}{\sqrt{n}}$, is also unknown.

3 In situations where the sample is small and the parent standard deviation or variance is unknown, there is little more that can be done unless you can assume that the parent population is normal. (In this case that is a reasonable assumption, the bats being a naturally occurring population.) If you can assume normality, then you may use the *t*-distribution, estimating the value of σ from your sample.

4 It is possible to test whether a set of data could reasonably have been taken from a normal distribution by using normal probability graph paper. The method involves making a cumulative frequency table and plotting points on a graph with specially chosen axes. If the graph obtained is approximately a straight line, then the data could plausibly have been drawn from a normal population. Otherwise a normal population is unlikely.

The *t*-distribution looks very like the normal distribution. Its exact shape depends on the number of degrees of freedom, ν, and, indeed, for large values of ν it is little different from it. The larger the value of ν, the closer the *t*-distribution is to the normal. Figure 2.1 shows the normal distribution and *t*-distributions $\nu = 2$ and $\nu = 10$.

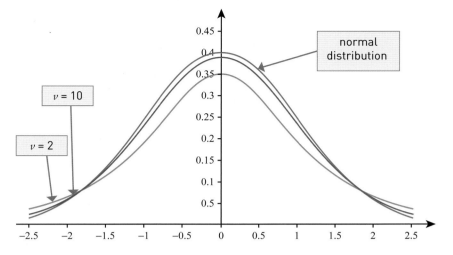

▲ Figure 2.1

Historical note

William S. Gosset was born in Canterbury, UK in 1876. After studying both mathematics and chemistry at Oxford, he joined the Guinness breweries in Dublin as a scientist. He found that an immense amount of statistical data was available, relating the brewing methods and the quality of the ingredients, particularly barley and hops, to the finished product. Much of this data took the form of samples, and Gosset developed techniques to handle them, including the discovery of the *t*-distribution. Gosset published his work under the pseudonym 'Student' and so the *t*-test is often called Student's *t*-test.

Gosset's name has frequently been misspelt as Gossett (with a double t), giving rise to puns about the *t*-distribution.

Confidence intervals using the *t*-distribution are constructed in much the same way as those using the normal, with the confidence limits given by

$$\bar{x} \pm k \frac{s}{\sqrt{n}}$$

where the values of k are found from tables. For instance, to find the value of k at the 5% significance level with 7 degrees of freedom, you need to look across the row $v = 7$ and the column $p = 0.975$ on the table given in Figure 2.2b on the next page.

If T has a *t*-distribution with v degrees of freedom then, for each pair of values of p and v, the table gives the value of t such that

$$P(T \leqslant t) = p.$$

▲ Figure 2.2a

Answers to exercises are available at www.hoddereducation.com/cambridgeextras

p	0.75	0.90	0.95	0.975	0.99	0.995	0.9975	0.999	0.9995
$v = 1$	1.000	3.078	6.314	12.71	31.82	63.66	127.3	318.3	636.6
2	0.816	1.886	2.920	4.303	6.965	9.925	14.09	22.33	31.60
3	0.765	1.638	2.353	3.182	4.541	5.841	7.453	10.21	12.92
4	0.741	1.533	2.132	2.776	3.747	4.604	5.598	7.173	8.610
5	0.727	1.476	2.015	2.571	3.365	4.032	4.773	5.894	6.869
6	0.718	1.440	1.943	2.447	3.143	3.707	4.317	5.208	5.959
7	0.711	1.415	1.895	2.365	2.998	3.499	4.029	4.785	5.408
8	0.706	1.397	1.860	2.306	2.896	3.355	3.833	4.501	5.041
9	0.703	1.383	1.833	2.262	2.821	3.250	3.690	4.297	4.781
10	0.700	1.372	1.812	2.228	2.764	3.169	3.581	4.144	4.587
11	0.697	1.363	1.796	2.201	2.718	3.106	3.497	4.025	4.437
12	0.695	1.356	1.782	2.179	2.681	3.055	3.428	3.930	4.318
13	0.694	1.350	1.771	2.160	2.650	3.012	3.372	3.852	4.221
14	0.692	1.345	1.761	2.145	2.624	2.977	3.326	3.787	4.140
15	0.691	1.341	1.753	2.131	2.602	2.947	3.286	3.733	4.073
16	0.690	1.337	1.746	2.120	2.583	2.921	3.252	3.686	4.015
17	0.689	1.333	1.740	2.110	2.567	2.898	3.222	3.646	3.965
18	0.688	1.330	1.734	2.101	2.552	2.878	3.197	3.610	3.922
19	0.688	1.328	1.729	2.093	2.539	2.861	3.174	3.579	3.883
20	0.687	1.325	1.725	2.086	2.528	2.845	3.153	3.552	3.850
21	0.686	1.323	1.721	2.080	2.518	2.831	3.135	3.527	3.819
22	0.686	1.321	1.717	2.074	2.508	2.819	3.119	3.505	3.792
23	0.685	1.319	1.714	2.069	2.500	2.807	3.104	3.485	3.768
24	0.685	1.318	1.711	2.064	2.492	2.797	3.091	3.467	3.745
25	0.684	1.316	1.708	2.060	2.485	2.787	3.078	3.450	3.725
26	0.684	1.315	1.706	2.056	2.479	2.779	3.067	3.435	3.707
27	0.684	1.314	1.703	2.052	2.473	2.771	3.057	3.421	3.689
28	0.683	1.313	1.701	2.048	2.467	2.763	3.047	3.408	3.674
29	0.683	1.311	1.699	2.045	2.462	2.756	3.038	3.396	3.660
30	0.683	1.310	1.697	2.042	2.457	2.750	3.030	3.385	3.646
40	0.681	1.303	1.684	2.021	2.423	2.704	2.971	3.307	3.551
60	0.679	1.296	1.671	2.000	2.390	2.660	2.915	3.232	3.460
120	0.677	1.289	1.658	1.980	2.358	2.617	2.860	3.160	3.373
∞	0.674	1.282	1.645	1.960	2.326	2.576	2.807	3.090	3.291

▲ **Figure 2.2b** Critical values for the t-distribution

You should be aware that values given in these statistical tables are rounded. Consequently the final digit of an answer that you obtain using these tables may not be quite the same as the answer you would obtain using the statistical function on your calculator instead. Most tables give 4 or 5 figures so it is good practice to round your final answer to 3 significant figures.

At the 5% significance level, you need 2.5% at each end of the distribution, so the value of p will be $100\% - \frac{1}{2} \times 5\% = 97.5\% = 0.975$.

To construct a 95% confidence interval for the mean weight of the bats, you look under $p = 0.975$ and $v = 7$, to get $k = 2.365$; see Figure 2.2b. This gives a 95% confidence interval of

$$146 - 2.365 \times \frac{9.27}{\sqrt{8}} \text{ to } 146 + 2.365 \times \frac{9.27}{\sqrt{8}}$$

138.2 to 153.8.

Another bat expert suggests that these bats are not, in fact, a new species, but from a known species. The average weight of adult males of this species is 160 grams. However, because the maximum value in the confidence interval is less than 160, in fact, only 153.8, this suggests that the expert may not be correct. Even if you use a 99% confidence interval, the upper limit is

$$146 + 3.499 \times \frac{9.27}{\sqrt{8}} = 157.5.$$

Therefore, it seems very unlikely that these bats are of the same species, based simply on their weights.

Example 2.1

A bus company is about to start a scheduled service between two towns some distance apart. Before deciding on an appropriate timetable, they do nine trial runs to see how long the journey takes. The times, in minutes, are

89	92	95	94	88	90	92	93	91

(i) Use these data to set up a 95% confidence interval for the mean journey time. You should assume that the journey times are normally distributed.

The company regards its main competition as the railway service, which takes 95 minutes.

(ii) Does your confidence interval provide evidence that the journey time by bus is different from that by train?

Solution

(i) For the given data,

$$n = 9, \qquad v = 9 - 1 = 8, \qquad \bar{x} = 91.56, \qquad s = 2.297.$$

For a 95% confidence interval, with $v = 8$, $k = 2.306$ (from tables). The confidence limits are given by

$$\bar{x} \pm k \frac{s}{\sqrt{n}} = 91.56 \pm 2.306 \times \frac{2.297}{\sqrt{9}}.$$

So the 95% confidence interval for μ is 89.79 to 93.33. ➡

Answers to exercises are available at www.hoddereducation.com/cambridgeextras

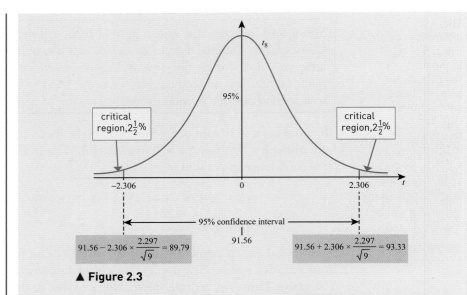

▲ Figure 2.3

(ii)   The confidence interval does not contain 95 minutes (the time taken by the train). Therefore there is sufficient evidence to suggest that the journey time by bus is different from that by train, and that it is, in fact, less.

2.2 Using the t-distribution for paired samples

The ideas developed in the last few pages can also be used in constructing confidence intervals for the difference in the means of paired data. This is shown in the next example.

Example 2.2

In an experiment on group behaviour, 12 subjects were each asked to hold one arm out horizontally while supporting a 2 kg weight, under two conditions:

» while together in a group

» while alone with the experimenter.

The times, in seconds, for which they were able to support the weight under the two conditions were recorded as follows.

Subject	A	B	C	D	E	F	G	H	I	J	K	L
'Group' time	61	71	72	53	71	43	85	72	82	54	70	73
'Alone' time	43	72	81	35	56	39	63	66	38	60	74	52
Difference	18	−1	−9	18	15	4	22	6	44	−6	−4	21

The 'true' difference means the difference in the times in the population as a whole. You cannot know with certainty the differences in the population but you can infer from the sample a confidence interval for the population.

Find a 90% confidence interval for the true difference between 'group' times and 'alone' times. You may assume that the differences are normally distributed. Does your result provide evidence that there is any difference in the times in the population as a whole?

Solution

The sample comprises the 12 differences.

Mean	$\bar{d} = 10.67$
Standard deviation	$s = 15.24$
Degrees of freedom	$v = 12 - 1 = 11$

Given the assumption that the differences are normally distributed, you may use the t-distribution.

For $v = 11$, the two-tailed critical value from the t-distribution at the 10% level of significance is 1.796.

The 90% symmetrical confidence interval for the mean difference between the 'group' and 'alone' times is

$$\bar{d} - 1.796 \times \frac{s}{\sqrt{12}} \text{ to } \bar{d} + 1.796 \times \frac{s}{\sqrt{12}}$$

2.77 to 18.57.

Since the confidence interval does not contain zero, there is evidence that there is a difference in the times in the population as a whole.

2.3 Hypothesis testing on a sample mean using the t-distribution

At the beginning of this chapter, you met the t-distribution. This is the distribution of a sample mean when the parent population is normally distributed but the standard deviation of the parent population is unknown and has to be estimated using the sample standard deviation, s. In addition to finding a confidence interval, you can also carry out a **t-test** (a hypothesis test based on the t-distribution).

Example 2.3

Tests are being carried out on a new drug designed to relieve the symptoms of the common cold. One of the tests is to investigate whether the drug has any effect on the number of hours that people sleep.

The drug is given in tablet form one evening to a random sample of 16 people who have colds. The number of hours they sleep may be assumed to be normally distributed and is recorded as follows.

→

Answers to exercises are available at www.hoddereducation.com/cambridgeextras

| 8.1 | 6.7 | 3.3 | 7.2 | 8.1 | 9.2 | 6.0 | 7.4 |
| 6.4 | 6.9 | 7.0 | 7.8 | 6.7 | 7.2 | 7.6 | 7.9 |

There is also a large control group of people who have colds but are not given the drug. The mean number of hours they sleep is 6.6.

(i) Use these data to set up a 95% confidence interval for the mean length of time somebody with a cold sleeps after taking the tablet.

(ii) Carry out a test, at the 1% significance level, of the hypothesis that the new drug has an effect on the number of hours a person sleeps.

Solution

(i) For the given data, $n = 16$, $v = 16 - 1 = 15$, $\bar{x} = 7.094$, $s = 1.276$.

For a 95% confidence interval, with $v = 15$, $k = 2.131$ (from tables).

The confidence limits are given by $\bar{x} \pm k\dfrac{s}{\sqrt{n}} = 7.094 \pm 2.131 \times \dfrac{1.276}{\sqrt{16}}$.

So the 95% confidence interval for μ is 6.41 to 7.77.

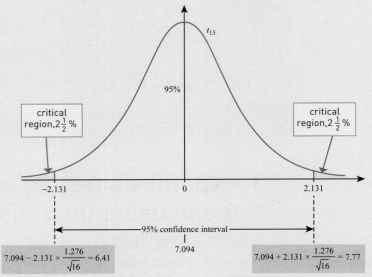

▲ **Figure 2.4**

(ii) H_0: There is no change in the mean number of hours sleep. $\mu = 6.6$

H_1: There is a change in the mean number of hours sleep. $\mu \neq 6.6$

Two-tailed test at the 1% significance level.

For this sample, $n = 16$, $v = 16 - 1 = 15$, $\bar{x} = 7.094$, $s = 1.276$.

The critical value for t, for $v = 15$, at the 1% significance level, is found from tables to be 2.947.

The test statistic $t = \left| \dfrac{\bar{x} - \mu}{\dfrac{s}{\sqrt{n}}} \right| = \left| \dfrac{7.094 - 6.6}{\dfrac{1.276}{\sqrt{16}}} \right| = 1.55.$

This is to be compared with 2.947, the critical value, for the 1% significance level.

Since 1.55 < 2.947, there is no reason at the 1% significance level to reject the null hypothesis.

There is insufficient evidence to suggest that the mean number of hours sleep is different when people take the drug.

Technology note

If you have access to a computer, you can use a spreadsheet (see Figure 2.5) to do all of the calculations for this test using the following steps.

1 Enter the data (in this case into cells B2 to B17).

2 Use the spreadsheet functions provided by your spreadsheet, for example =AVERAGE and =STDEV to find the mean and sample standard deviation.

3 Calculate the t-value using the formula =(B18−6.6)/(B19/SQRT(16)).

4 Use the spreadsheet function provided by your spreadsheet, for example =T.INV.2T(0.01,15) to calculate the critical value.

5 You can also find the p-value using the spreadsheet function provided by your spreadsheet, for example =T.DIST.2T(1.5482,15).

	A	B
1		Data
2		8.1
3		6.7
4		3.3
5		7.2
6		8.1
7		9.2
8		6.0
9		7.4
10		6.4
11		6.9
12		7.0
13		7.8
14		6.7
15		7.2
16		7.6
17		7.9
18	Mean	7.0938
19	Sample sd	1.2757
20	n	16
21	t value	1.5482
22	Critical t	2.9467
23	p-value	0.1424

In this case, the values of $\bar{x}$ and s are given in cells B18 and B19.

▲ Figure 2.5

---

1 The mean of a random sample of seven observations of a normally distributed random variable X is 132.6. Based on these seven observations, an unbiased estimate of the parent population variance s^2 is 148.84.

   (i) Explain why an estimate of the standard error is given by 4.61.

   (ii) Show that a 95% confidence interval for the mean μ of X is 121.3 to 143.9.

2 The weights in grams of six beetles of a particular species are as follows.

   12.3   9.7   11.8   10.1   11.2   12.4

   (i) Calculate the sample mean and show that an estimate of the sample variance is 1.291.

   (ii) Show that a 90% confidence interval for the mean μ of X is 10.32 to 12.18.

Answers to exercises are available at www.hoddereducation.com/cambridgeextras

M **3** An aptitude test for entrance to university is designed to produce scores that may be modelled by the normal distribution. In early testing, 15 students from the appropriate age group are given the test. Their scores (out of 500) are as follows.

321	445	219	378	317	407	289	345
276	463	265	165	340	298	315	

(i)   Use these data to estimate the mean and standard deviation to be expected for students taking this test.

(ii) Construct a 95% confidence interval for the mean.

M **4** A fruit farmer has a large number of almond trees, all of the same variety and of the same age. One year, he wishes to estimate the mean yield of his trees. He collects all the almonds from eight trees and records the following weights (in kg).

36	53	78	67	92	77	59	66

(i)   Use these data to estimate the mean and standard deviation of the yields of all the farmer's trees.

(ii) Construct a 95% confidence interval for the mean yield.

(iii) What statistical assumption is required for your procedure to be valid?

(iv) How might you select a sample of eight trees from those growing in a large field?

M **5** A forensic scientist is trying to decide whether a man accused of fraud could have written a particular letter. As part of the investigation she looks at the lengths of sentences used in the letter. She finds them to have the following numbers of words.

17	18	25	14	18	16	14	16	16	21	25	19

(i)   Use these data to estimate the mean and standard deviation of the lengths of sentences used by the letter writer.

(ii) Construct a 90% confidence interval for the mean length of the letter writer's sentences.

(iii) What assumptions have you made to obtain your answer?

(iv) A sample of sentences written by the accused has mean length 26 words. Does this mean he is innocent?

M **6** A large company is investigating the number of incoming telephone calls at its exchange, in order to determine how many telephone lines it should have. On Sundays very few calls are received because the office is closed. During March one year, the number of calls received each day was recorded, as follows.

623	584	598	701	656	210	23	655	661	599
634	681	197	25	592	643	642	698	659	201
19	588	672	612	706	650	212	29	681	642
677									

(i) What day of the week was 1 March?

(ii) Which of the data do you consider relevant to the company's research and why?

(iii) Construct a 95% confidence interval for the number of incoming calls per weekday.

(iv) Your calculation is criticised on the grounds that your data are discrete and so the underlying distribution cannot possibly be normal. How would you respond to this criticism?

(M) 7 A tyre company is trying out a new tread pattern, which it is hoped will result in the tyres giving greater distance. In a pilot experiment, 12 tyres are tested; the mileages ($\times 1000$ miles) at which they are condemned are as follows:

65   63   71   78   65   69   59   81   72   66   63   62

(i) Construct a 95% confidence interval for the mean distance that a tyre travels before being condemned.

(ii) What assumptions, statistical and practical, are required for your answer to part (i) to be valid?

(M) 8 A large fishing-boat made a catch of 500 mackerel from a shoal. The total mass of the catch was 320 kg. The standard deviation of the mass of individual mackerel is known to be 0.06 kg.

(i) Find a 99% confidence interval for the mean mass of a mackerel in the shoal.

An individual fisherman caught ten mackerel from the same shoal. These had masses (in kg) of:

1.04   0.94   0.92   0.85   0.85   0.70   0.68   0.62   0.61   0.59

(ii) From these data only, use your calculator to estimate the mean and standard deviation of the masses of mackerel in the shoal.

(iii) Assuming the masses of mackerel are normally distributed, use your results from part (ii) to find another 99% confidence interval for the mean mass of a mackerel in the shoal.

(iv) Give two statistical reasons why you would use the first limits you calculated in preference to the second limits.

Answers to exercises are available at www.hoddereducation.com/cambridgeextras

M | **9** A new computerised job-matching system has been developed that finds suitably skilled applicants to fit notified vacancies. It is hoped that this will reduce unemployment rates, and a trial of the system is conducted in seven areas.

The unemployment rates in each area just before the introduction of the system and after one month of its operation are recorded in the table below.

Area	1	2	3	4	5	6	7
Rate before new system (%)	10.3	3.6	17.8	5.1	4.6	11.2	7.7
Rate after new system (%)	9.3	4.1	15.2	5.0	3.3	10.3	8.1

(i) Find a 90% confidence interval for the true difference between the two rates of unemployment.

(ii) Does your confidence interval provide evidence that there is a difference in the rates after the new system is introduced?

(iii) Do you think the assumptions required to construct the confidence interval are justified here?

M | **10** Two timekeepers at an athletics track are being compared. They each time the nine sprints one afternoon.

(i) Find a 99% confidence interval for the true difference between the times recorded by the two timers. The times they record are listed below.

Race	1	2	3	4	5	6	7	8	9
Timer 1	9.65	10.01	9.62	21.90	20.70	20.90	42.30	43.91	43.96
Timer 2	9.66	9.99	9.44	22.00	20.82	20.58	42.39	44.27	44.22

(ii) Do you think that the two timers are equivalent on average?

(iii) Are the assumptions appropriate for a confidence interval based on the *t*-test justified in this case?

M | **11** Fourteen marked rats were timed twice as they ran through a maze. In one condition, they had just been fed; in the other they were hungry.

(i) Find a 95% confidence interval for the true difference between the rats' times when they are fed and when they are hungry. The data below give the rats' times in each condition.

Rat	A	B	C	D	E	F	G	H	I	J	K	L	M	N
Fed time (seconds)	30	31	25	23	50	26	14	27	31	39	38	39	44	30
Hungry time (seconds)	29	18	14	27	37	34	15	22	29	18	20	10	30	32

(ii) Do you think the assumptions required to construct the confidence interval are justified here?

(iii) Half of the rats were made to run the maze first when hungry and half ran it first when fed. Why did the experimenter do this?

12 Cans of cola are supposed to contain at least 330 ml. Salma thinks that the average content is less than this. She measures the contents of a random sample of ten cans with the following results (all measured in ml).

326.5	331.2	327.9	329.8	330.4
327.6	329.3	331.0	328.4	330.1

(i) What distributional assumption is necessary in order to carry out a *t*-test to check Salma's suspicion?

(ii) Carry out this test at the 5% significance level.

(iii) If Salma picked 10 cans from a single box of 24, would the test have still been valid? Explain your answer.

13 In a certain country, past research showed that in the average married couple, the man was 7 cm taller than his wife. A sociologist believes that, with changing roles, people are now choosing marriage partners nearer their own height. She has measured the heights of 12 couples. Her results are shown in the table below.

Couple	1	2	3	4	5	6
Man	188.2	174.2	192.4	163.4	183.2	171.4
Woman	178.4	165.1	191.9	156.3	178.7	163.0
Couple	7	8	9	10	11	12
Man	180.6	173.5	166.8	171.9	175.5	163.2
Woman	180.4	170.2	164.2	166.2	176.9	158.8

(i) Test at the 5% significance level the hypothesis that men are on average less than 7 cm taller than their wives.

(ii) Explain clearly what assumptions you make in this case.

14 The speed v at which a javelin is thrown by an athlete is measured in km h^{-1}. The results for 10 randomly chosen throws are summarised by

$$\sum v = 1110.8, \qquad \sum (v - \bar{v})^2 = 333.9,$$

where $\bar{v}$ is the sample mean.

(i) Stating any necessary assumption, calculate a 99% confidence interval for the mean speed of a throw.

The results for a further 5 randomly chosen throws are now combined with the above results. It is found that the sample variance is smaller than that used in part (i).

(ii) State, with reasons, whether a 95% confidence interval calculated from the combined 15 results will be wider or less wide than that found in part (i).

Cambridge International AS & A Level Further Mathematics
9231 Paper 23 Q7 November 2012

Answers to exercises are available at www.hoddereducation.com/cambridgeextras

15 In a crossword competition the times, x minutes, taken by a random sample of 6 entrants to complete a crossword are summarised as follows.

$$\sum x = 210.9 \qquad \sum (x - \bar{x})^2 = 151.2$$

The time to complete a crossword has a normal distribution with mean μ minutes. Calculate a 95% confidence interval for μ.

Assume now that the standard deviation of the population is known to be 5.6 minutes. Find the smallest sample size that would lead to a 95% confidence interval for μ of width at most 5 minutes.

Cambridge International AS & A Level Further Mathematics
9231 Paper 21 Q8 June 2011

16 A random sample of 8 observations of a normal random variable X gave the following summarised data, where $\bar{x}$ denotes the sample mean.

$$\sum x = 42.5 \qquad \sum (x - \bar{x})^2 = 15.519$$

Test, at the 5% significance level, whether the population mean of X is greater than 4.5.

Calculate a 95% confidence interval for the population mean of X.

Cambridge International AS & A Level Further Mathematics
9231 Paper 21 Q9 June 2012

17 A random sample of 10 observations of a normally distributed random variable X gave the following summarised data, where $\bar{x}$ denotes the sample mean.

$$\sum x = 70.4 \qquad \sum (x - \bar{x})^2 = 8.48$$

Test, at the 10% significance level, whether the population mean of X is less than 7.5.

Cambridge International AS & A Level Further Mathematics
9231 Paper 21 Q7 November 2013

18 A random sample of 8 sunflower plants is taken from the large number grown by a gardener, and the heights of the plants are measured. A 95% confidence interval for the population mean, μ metres, is calculated from the sample data as $1.17 < \mu < 2.03$. Given that the height of a sunflower plant is denoted by x metres, find the values of $\sum x$ and $\sum x^2$ for this sample of 8 plants.

Cambridge International AS & A Level Further Mathematics
9231 Paper 21 Q7 June 2015

2.4 Using the *t*-distribution with two samples

Have you noticed how time often seems to pass more slowly after lunch?

If time passes more slowly, one minute of real time should seem longer, so if you ask people to estimate when a minute appears to have elapsed, the real time elapsed will be less.

You could ask the question: 'Will the mean real time elapsed when one minute appears to have elapsed be less after lunch than before?'

In this example, you are interested, not in what the mean value of a random variable is, but in what the difference between the mean values is in two different situations. Statistical problems giving rise to different versions of this general question are the topic of the next few sections.

INVESTIGATION

Lunch time

Find a group of volunteers and approach them before lunch. Give each of them a starting signal and ask to say when one minute has elapsed. Record the real time elapsed. You will then need to find a second group of volunteers to approach after lunch. You will need reasonably large groups to get useful results.

1 What are the advantages and disadvantages of using separate groups of people for the before-lunch and after-lunch times?

2 What are the advantages and disadvantages of instead conducting an experiment in which the same people are asked before and after lunch and only the difference in their real times recorded?

The volunteers in research projects are called **subjects** and 'before lunch' and 'after lunch' are the two **conditions** in which testing occurs. An experiment such as the one described above, where a different group of subjects is tested in each of the two conditions is called an **unpaired design**; this is in contrast to a **paired design** you met earlier in this chapter, where the same set of subjects is tested in both conditions.

The members of a maths class were asked one morning to check the time shown by their watches, then look away and, when they estimated that a minute had elapsed, to check their watches again to see how long had in fact elapsed.

The same procedure was followed with another class, from the same year group, that afternoon. The back-to-back stem-and-leaf diagram on the next page shows the results.

Answers to exercises are available at www.hoddereducation.com/cambridgeextras

Morning class		Afternoon class
(24 students)		(22 students)
	2	8
2	3	1
5 5 6	3	6
	4	4 4 0
7 8	4	9 9 9 6 6 5
1 1 3 3 3 3 4 4	5	2 2 1 1 1
5 5 5 7 7 7 7	5	8 7 7 5
1 4 4	6	3

$6 \mid 3$ represents 63 seconds

▲ **Figure 2.6**

This experiment gives a set of data with which you could investigate the question asked at the start of this section. This experiment has an unpaired design: two separate groups of subjects are used in the two conditions.

This section uses the data given in Figure 2.6 to work through the process of hypothesis testing in the context of an unpaired design. If you have carried out your own investigation, you might find it helpful to repeat the calculations using your data.

You cannot look here at the difference between a before-lunch and an after-lunch time for a particular person, but you can look at the difference between the mean before-lunch time and the mean after-lunch time. In fact, you can make the following hypotheses:

H_0: There is no difference between the mean of people's estimates of one minute before and after lunch.

H_1: After lunch, the mean of people's estimates of one minute tends to be shorter than before lunch.

You can then use as your sample statistic the difference between the before-lunch sample mean and the after-lunch sample mean. You need to calculate the distribution of this sample statistic on the assumption that the null hypothesis is true.

The test statistic and its distribution

Assume that each before-lunch estimate is an independent random variable X_i $(i = 1, \ldots, 24)$ with the normal distribution $N(\mu_b, \sigma_b^2)$ and each after-lunch estimate is an independent random variable Y_j $(j = 1, \ldots, 22)$ with normal distribution $N(\mu_a, \sigma_a^2)$. You are also making the assumption that each X is independent of each Y. This is a plausible assumption; it merely requires the independence of the two samples taken.

Recall that the mean of a sample size n from a normal distribution $N(\mu, \sigma^2)$ has distribution

$$N\left(\mu, \frac{\sigma^2}{n}\right).$$

In this case, therefore, the mean of the 24 before–lunch estimates has distribution

$$\bar{X} \sim N\left(\mu_b, \frac{\sigma_b^2}{24}\right)$$

and the mean of the 22 after-lunch estimates has distribution

$$\bar{Y} \sim N\left(\mu_a, \frac{\sigma_a^2}{22}\right).$$

Next you need a result that if X has distribution $N(\mu_x, \sigma_x^2)$ and Y has distribution $N(\mu_y, \sigma_y^2)$ then $(X - Y)$ has distribution

$$N\left(\mu_X - \mu_Y, \sigma_X^2 + \sigma_Y^2\right).$$

Here, the distribution of the differences of the two sample means is therefore

$$\bar{X} - \bar{Y} \sim N\left(\mu_b - \mu_a, \frac{\sigma_b^2}{24} + \frac{\sigma_a^2}{22}\right).$$

The null hypothesis then states that both means are equal, i.e. $\mu_b = \mu_a$ and so, if the null hypothesis is true

$$\bar{X} - \bar{Y} \sim N\left(0, \frac{\sigma_b^2}{24} + \frac{\sigma_a^2}{22}\right)$$

Unfortunately, you do not know σ_b^2 or σ_a^2, so you will want, as you have in earlier work, to replace these unknown values with sample estimates. It might seem most natural to use sample estimates for the two unknown variances, but in fact in turns out that it is then hard to make any progress in calculating the distribution. This chapter, therefore, only deals with the case where you can assume that the variances in the two conditions are equal: here, this means that $\sigma_b^2 = \sigma_a^2 = \sigma^2$; that is, the before- and after-lunch estimates have the same variance. In this case

$$\bar{X} - \bar{Y} \sim N\left(0, \sigma^2 \left(\frac{1}{24} + \frac{1}{22}\right)\right).$$

So that

$$\frac{\bar{X} - \bar{Y}}{\sqrt{\sigma^2\left(\frac{1}{24} + \frac{1}{22}\right)}} = \frac{\bar{X} - \bar{Y}}{\sigma\sqrt{\frac{1}{24} + \frac{1}{22}}} \sim N(0, 1). \qquad \text{①}$$

To estimate σ^2, you can use the pooled variance estimator from the two samples,

$$S^2 = \frac{(24 - 1)S_b^2 + (22 - 1)S_a^2}{(24 + 22 - 2)}$$

Answers to exercises are available at www.hoddereducation.com/cambridgeextras

where S_b^2 and S_a^2 are the usual unbiased sample estimators of the population variance given by

$$S_b^2 = \frac{\sum_{i=1}^{24}(X_i - \overline{X})^2}{23} = \frac{\sum_{i=1}^{24} X_i^2 - 24\overline{X}^2}{23} \quad \text{and} \quad S_a^2 = \frac{\sum_{i=1}^{22}(Y_i - \overline{Y})^2}{21} = \frac{\sum_{i=1}^{22} Y_i^2 - 22\overline{Y}^2}{21}.$$

The test statistic,

$$\frac{\overline{X} - \overline{Y}}{S\sqrt{\frac{1}{24} + \frac{1}{22}}}$$

which is obtained from ① by replacing the value of σ^2 with its estimator S^2, then has a *t*-distribution, with degrees of freedom equal to that in the pooled variance estimate: $(24 + 22 - 2) = 44$.

Carrying out the *t*-test for an unpaired sample

In the example

$$\overline{x} = 51.542, \; s_b = 8.797; \; \overline{y} = 47.909, \; s_a = 8.574$$

so

$$s = \sqrt{\frac{23 \times 8.797^2 + 21 \times 8.574^2}{44}} = 8.691$$

and the value of the test statistic is

$$\frac{51.542 - 47.909}{8.691 \times \sqrt{\frac{1}{24} + \frac{1}{22}}} = 1.416.$$

The critical region for a one-tailed test, in the case of 44 degrees of freedom, at the 5% significance level is $t > 1.680$ (this value does not appear in the tables, but can be obtained by interpolation for the values given for 30 and 50 degrees of freedom). Since $1.416 < 1.680$, the results lead you to accept the null hypothesis at this significance level: there is no good evidence that the before-lunch mean and the after-lunch mean of the population as a whole are different.

> **Note**
>
> The process described is a *t*-test for the difference of two means with unpaired samples, also known as a 2-sample *t*-test.

INVESTIGATION

Carry out at least one of the experiments below, and test the hypothesis given using a 2-sample *t*-test.

1   Use a reaction timer to decide whether males and females have the same mean reaction times, or whether older people have slower reactions than young people (you can choose the definition of 'older' to suit the samples you have available) or whether squash players have quicker reactions than non-players.

You can use a 30 cm ruler as a reaction timer: hold the ruler vertically with the zero mark downwards, while the subject holds their thumb and forefinger 2 cm apart at the zero mark of the ruler. You drop the ruler without warning and your subject tries to catch it between thumb and forefinger. The distance, d, in millimetres, through which the ruler has fallen before it is caught can be used to measure the reaction time, t, in seconds, using the formula

$$t = \frac{\sqrt{d}}{70}.$$

2   Are students studying A Level maths better at mental arithmetic than those taking other A Levels? You will need to devise a mental arithmetic test (do you want to test speed or accuracy?) and administer it to a group of A Level maths students and a group of students taking other A Levels. *Do not be disappointed by the results!* You can adapt this test to suit your prejudices: are A Level geography students better at naming capitals of foreign countries? Are A Level English students better at spelling?

3   Two groups of subjects are each given lists of 25 words. Both groups must run down the list as quickly as possible. Those in the first group tick the words that are in capital letters. (You should make sure that about half of the words, placed randomly in the list, are in capital letters.) The second group ticks the words that rhyme with a target word that you give them. (Make sure that about half of the words, placed randomly in the list, do rhyme with this word.) You then ask the subjects each to write down as many words as they can remember from the list: do not tell the subjects in advance that they will have to do this. Test the hypothesis that the subjects who have looked for rhymes remember more of the words than those who looked for capital letters. Why would it be difficult to run this experiment with a paired design?

Assumptions for the 2-sample *t*-test

The assumptions needed for the 2-sample *t*-test are quite severe.

1   The two samples must be independent random samples of the populations involved.

   Strictly, this requires every possible sample to have an equal probability of being chosen. If you simply picked a group of volunteers, it would, therefore, probably not be a random sample. However, this method is very close to the method often used by academic psychologists when choosing their samples. The hope in choosing a random sample is that the effects of all the irrelevant differences between members of the population that influence the variables you are testing will average out.

2   The random variables measured in the two conditions must:

   (i)   be normally distributed

   (ii)   have equal variances in the two conditions.

Are these assumptions justified? The only information you have to help you decide is the two samples: the stem-and-leaf diagram for the data of the example is shown again below.

Morning class		**Afternoon class**
(24 students)		(22 students)
	2	8
2	3	1
5 5 6	3	6
	4	4 4 0
7 8	4	9 9 9 6 6 5
1 1 3 3 3 3 4 4	5	2 2 1 1 1
5 5 5 7 7 7 7	5	8 7 7 5
1 4 4	6	3

$\quad \mid 6 \mid 3 \quad$ represents 63 seconds

▲ **Figure 2.7**

At first sight, the distributions here do not look much like samples from a normal distribution: they are rather obviously negatively skewed. Neither is it clear that they would have come from populations of the same variance. However, these are relatively small samples and it would be unwise to draw any firm conclusions from them about the population distribution from which they are drawn.

> › Do you think the assumptions made in the 2-sample *t*-test are justified in the case of the experiment you carried out?

The underlying logic of hypothesis testing

When you construct the sampling distribution of a test statistic you use:

» a model for the distribution of the random variables involved in the statistic

» the value given to a parameter of this distribution by the null hypothesis.

In the time-estimation example, the construction of the sampling distribution depends on:

» people's estimates of one minute before and after lunch being independent and distributed normally, with a common variance

» the null hypothesis that the difference between the means of their estimates before and after lunch is zero.

The alternative hypothesis, that the difference between these means is greater than zero, gives an alternative range of possible values for the parameter of the distribution, but assumes the same model for the random variables involved.

In the example it was determined (by using pre-calculated tables, in fact), that if:

» the model for the random variables was correct

and

» the null hypothesis were true

then a test statistic greater than 1.680 would only arise in a random sample 5% of the time. (The significance level is 5%.)

In the example earlier in this section, you obtained a value of 1.416 and, since this is less than 1.680, the null hypothesis was accepted. Suppose, instead, that you had obtained a value greater than 1.680; say, for example, 1.832. In that case, there would be three possible explanations.

Explanation A

» The model is correct.
» The null hypothesis is false, because the mean difference in before- and after-lunch times is greater than zero.

Explanation B

» The model is correct.
» The null hypothesis is true (or false because the mean difference in before- and after-lunch times is actually less than zero).

However, the sample selected happens to give a value of the test statistic greater than 1.680. The probability of this happening is 0.05 (the significance level) if the null hypothesis is true, or less if the mean difference in before- and after-lunch times is actually less than zero.

Explanation C

» The model is incorrect, because the sampling method does not produce independent estimates for each subject, or because the estimates are not distributed normally in the population, or do not have a common variance.
» The null hypothesis is true or false.

In this, case you have no idea how likely it is that the test statistic will have any value at all.

The hypothesis testing methodology is:

» to assume that explanation C is not the case
» to observe that if explanation B was the case, then the results obtained would be very unlikely
» and therefore to accept that explanation A is the case.

Thus you reject the null hypothesis and accept the alternative.

However, you should always be aware that the logic that leads you to this conclusion on the basis of the evidence in the sample depends on the correctness of your sampling and distributional assumptions.

Answers to exercises are available at www.hoddereducation.com/cambridgeextras

2.5 Comparison between paired and 2-sample *t*-tests

The table below shows summarised data from the experiment you have just been analysing, together with data gathered from a paired experiment using a single sample of twelve people. Each was asked, both before and after lunch, to estimate one minute in the same way as described for the unpaired design.

	Data from paired experiment	Data from unpaired experiment
Before-lunch time	$\bar{x} = 51.250$, $s_b = 8.237$, $n = 12$	$\bar{x} = 51.542$, $s_b = 8.797$, $n = 24$
After-lunch time	$\bar{y} = 47.583$, $s_a = 9.258$, $n = 12$	$\bar{y} = 47.909$, $s_a = 8.574$, $n = 22$

The test statistic for the paired experiment is 1.829, with 11 degrees of freedom and a critical value of 1.796, so that here the null hypothesis is rejected.

Why do you reject the null hypothesis in the paired case where the sample size is considerably smaller, which, all other things being equal, would usually lead to a less decisive test, as reflected in the larger critical value?

You can see why the opposite appears to have happened if you look at how the test statistics for the two cases are calculated.

Test statistic for paired experiment	Test statistic for unpaired experiment
$\dfrac{51.250 - 47.583}{6.946\sqrt{\dfrac{1}{12}}} = 1.829$	$\dfrac{51.542 - 47.909}{8.691\sqrt{\dfrac{1}{24} + \dfrac{1}{22}}} = 1.416$

The test statistics for the paired and unpaired calculations have very similar numerators, but the standard error in the denominator is considerably larger in the unpaired calculation, despite the larger sample size in that case.

Note

The crucial point is that there is, for all sorts of reasons, considerable variation amongst people in their reaction times and lunch is only one, relatively small, effect amongst many. Some people will tend to make short estimates in both conditions and some long estimates in both conditions, though in both cases the effect of lunch may be the same.

The paired design enables you to take this into account in a way that the unpaired design cannot because of the way the standard error is estimated.

Using paired and 2-sample *t*-tests

It is a characteristic of research by social scientists that they are looking for a small average difference between the values of a particular variable in two different conditions, but that subjects show very substantial variation in the values of this variable within both conditions. In these situations, a 2-sample *t*-test is not usually very helpful, as it will require a very large sample size to discriminate between the null hypothesis of no difference between the means in the two conditions and the true situation where there is a small difference.

Considerable ingenuity is therefore employed in attempting to match subjects so that a paired test can be used to eliminate some of the variation between them and the small difference between the two experimental conditions is not swamped.

In the paired experiment, you used the same subject in each of two conditions, but this is not necessary. In fact, having taken part in one experimental condition sometimes makes it impossible to take part in the second.

For example, if you wish to test the effect on children's intelligence of an upbringing in families from two different social classes, you could not use the same child and bring it up twice, nor would a 2-sample *t*-test be suitable in this case: the variation in intelligence caused by other factors would swamp the effect you are looking for.

One possibility is to find pairs of identical twins who are being adopted at birth and are assigned to adoptive parents of different social classes: these constitute matched pairs of subjects and you could use a *t*-test on the differences between the intelligences of the twins from the two types of family. Notice that here the matching is perfect in the sense that both children have identical genetic endowments: the belief implicit in this experiment is that heredity is a major cause of variation in intelligence and this effect will be cancelled out by the matching process. Of course, there will be many differences between the adoptive families other than class, and it is possible that the variations in intelligence induced by these differences in upbringing will still swamp the effect being examined. Ideally, you would want to find identical twins being assigned to families differing only in their social class, but it is unlikely that you would find enough, if any, examples of this to conduct the test!

2.6 Testing for a non-zero value of the difference of two means

You have now used the *t*-test to examine the null hypothesis that two different conditions produce the same mean value of some random variable. The method can also be used in a more general way to test null hypotheses that suggest that the mean of a random variable, X, differs by a given amount in the two conditions.

Hypothesis: for some given value of δ

H_0: The difference between the mean values of X in condition 1 and condition 2 is δ.

Answers to exercises are available at www.hoddereducation.com/cambridgeextras

Sample: Two sets of observations of X, one set in each condition.

Let X_1 and X_2 be the random variables in the two conditions and n_1 and n_2 be the number of observations under each condition.

Use these values to calculate the sample means $\bar{X}_1$ and $\bar{X}_2$ and the unbiased pooled–sample estimator S^2 of the population variance.

Then

$$\frac{(\bar{X}_1 - \bar{X}_2) - \delta}{S\sqrt{\dfrac{1}{n_1} + \dfrac{1}{n_2}}} \sim t_{n_1 + n_2 - 2}$$

provided that the random variable X is distributed normally in the population, with the same variance in each condition, and that the null hypothesis is true.

Example 2.4

The manufacturers of a dieting compound claim that the use of their product as part of a calorie-counting diet leads to an average extra weight loss of at least five kilograms in a period of months. An experiment has been carried out by a consumers' group that doubts this claim.

The hypotheses are:

H_0: The mean extra weight loss in a period of months from adding the dieting compound to a calorie-counting diet is five kilograms.

H_1: The mean extra weight loss in a period of months from adding the dieting compound to a calorie-counting diet is less than five kilograms.

The assumptions are that the weight loss in a period of months from a calorie-counting diet, with or without the dieting compound, is a normally distributed random variable and that the addition of the dieting compound to the diet does not affect the variance of this random variable.

Thirty-six dieters used the dieting compound with their diets; their weight losses x_i ($i = 1, \ldots, 36$) in kilograms are summarised by the figures

$$\sum_{i=1}^{36} x_i = 409.32 \qquad \sum_{i=1}^{36} x_i^2 = 6102.39.$$

Sixty-two dieters followed the same calorie-counting procedure, but did not use the dieting compound; their weight losses y_j ($j = 1, \ldots, 62$) in kilograms are summarised by the figures

$$\sum_{j=1}^{62} y_j = 571.64 \qquad \sum_{j=1}^{62} y_j^2 = 5618.40.$$

Solution

These data give: $\bar{x} = 11.37$, $s_x = 6.433$, $\bar{y} = 9.22$ and $s_y = 2.388$

so that

$$s = \sqrt{\frac{35 \times 6.433^2 + 61 \times 2.388^2}{36 + 62 - 2}} = 4.326$$

The test statistic is

$$\frac{(11.37 - 9.22) - 5}{4.326 \times \sqrt{\frac{1}{36} + \frac{1}{62}}} = -3.144$$

and there are (size of sample 1 + size of sample 2 − 2) = (36 + 62 − 2) = 96 degrees of freedom.

The critical region for a one-tailed test with 96 degrees of freedom at the 5% significance level is $t < -1.661$ using interpolation and so, since $-3.144 < -1.661$, the null hypothesis is rejected in favour of the alternative that the average extra weight loss is not as great as five kilograms.

2.7 Hypothesis tests and confidence intervals

There is a very close relationship between hypothesis tests and confidence intervals, which should be clearly understood.

A **hypothesis test** suggests a value for an unknown population parameter (the null hypothesis), and then accepts this value if a test statistic lies in a particular range (that is, lies outside the critical region). However, the critical region depends on the hypothesised population parameter, so you can reverse this process. Thus, for a given value of the test statistic, you can determine the range of values for the population parameter that would be accepted by the test if they were offered as null hypotheses. This is called the **confidence interval** for the population parameter.

For instance, in the case where you take an independent random sample of size n from a normal distribution to test the hypotheses:

H_0: Population mean = μ

H_1: (a) Population mean ≠ μ

or (b) Population mean > μ

or (c) Population mean < μ

The test statistic is

$$\frac{\bar{x} - \mu}{\frac{s}{\sqrt{n}}}$$

and you accept the null hypothesis at the $a\%$ significance level if

(a) $-\tau_2 < \dfrac{\bar{x} - \mu}{\frac{s}{\sqrt{n}}} < \tau_2$ or   (b) $\dfrac{\bar{x} - \mu}{\frac{s}{\sqrt{n}}} < \tau_1$ or   (c) $\dfrac{\bar{x} - \mu}{\frac{s}{\sqrt{n}}} > -\tau_1$

where τ_1, τ_2 are the one- and two-tailed critical values respectively for the t-distribution with $n - 1$ degrees of freedom at the $a\%$ level.

Answers to exercises are available at www.hoddereducation.com/cambridgeextras

Alternatively, for a given value of $\bar{x}$ you can view these inequalities as constraining the range of values of μ, which would be accepted by the test if they were offered as null hypotheses, and rearranging them gives the $(100 - a)\%$ confidence intervals.

> ## Note
> Usually two-sided confidence intervals are used, as in (a).

(a) $\bar{x} - \tau_2 \dfrac{s}{\sqrt{n}} < \mu < \bar{x} + \tau_2 \dfrac{s}{\sqrt{n}}$ or (b) $\bar{x} - \tau_1 \dfrac{s}{\sqrt{n}} < \mu$ or

(c) $\bar{x} + \tau_1 \dfrac{s}{\sqrt{n}} > \mu$

Confidence intervals for the difference of two means from unpaired samples

Two runners are being considered for a place in a team. They have each recently competed in several races, though not against each other. Their times (in seconds) were as shown in the table below.

Runner 1	47.2	51.8	48.1	47.9	49.0	48.2	48.1
Runner 2	49.5	47.4	48.3	49.1	47.6		

You can model the first and second runners' times with variables T_1 and T_2 with distributions $N(\mu_1, \sigma^2)$ and $N(\mu_2, \sigma^2)$, respectively. You are describing their running times as normally distributed with different means and a common variance. The different means reflect differences in the runners' underlying ability; the random variability comes from factors such as the influence of other runners and weather conditions for which the effects in the different races are independent.

Because you are interested in the difference in the runners' underlying abilities, you are looking for a confidence interval for the difference between μ_1 and μ_2.

The sample means of the runners' times have distributions

$$\bar{T}_1 \sim N\left(\mu_1, \tfrac{1}{7}\sigma^2\right) \text{and } \bar{T}_2 \sim N\left(\mu_2, \tfrac{1}{5}\sigma^2\right)$$

so that the distribution of their differences is

$$(\bar{T}_1 - \bar{T}_2) \sim N\left(\mu_1 - \mu_2, \sigma^2\left(\tfrac{1}{7} + \tfrac{1}{5}\right)\right).$$

The standardised variable

$$\frac{(\bar{T}_1 - \bar{T}_2) - (\mu_1 - \mu_2)}{\sigma\sqrt{\left(\tfrac{1}{7} + \tfrac{1}{5}\right)}}$$

then has an $N(0, 1)$ distribution.

If you replace σ^2 with its unbiased sample estimator,

$$S^2 = \frac{(7-1)S_1^2 + (5-1)S_2^2}{(7+5-2)}$$ where S_1^2 and S_2^2 are the unbiased sample

estimators of the variance from the two separate samples, then, finally

$$D = \frac{(\overline{T}_1 - \overline{T}_2) - (\mu_1 - \mu_2)}{S\sqrt{\left(\frac{1}{7} + \frac{1}{5}\right)}}$$

has distribution $t_{7+5-2} = t_{10}$.

The critical value for the t-distribution with 10 degrees of freedom at the 5% significance level is 2.228, so that D lies between -2.228 and $+2.228$ in 95% of samples; that is, a 95% confidence level for $(\mu_1 - \mu_2)$ is defined by:

$$-2.228 < \frac{(\overline{t}_1 - \overline{t}_2) - (\mu_1 - \mu_2)}{s\sqrt{\left(\frac{1}{7} + \frac{1}{5}\right)}} < +2.228.$$

This can be rearranged as

$$(\overline{t}_1 - \overline{t}_2) - 2.228s\sqrt{\left(\frac{1}{7} + \frac{1}{5}\right)} < (\mu_1 - \mu_2) < (\overline{t}_1 - \overline{t}_2) + 2.228s\sqrt{\left(\frac{1}{7} + \frac{1}{5}\right)}.$$

For the data here,

$$\overline{t}_1 = 48.614, \qquad s_1^2 = 2.2514$$

$$\overline{t}_2 = 48.380, \qquad s_2^2 = 0.8370$$

so that

$$s^2 = \frac{(7-1)s_1^2 + (5-1)s_2^2}{(7+5-2)} = 1.686.$$

Thus the confidence level is

$$-1.460 < (\mu_1 - \mu_2) < 1.928.$$

Note

Note the considerable width of this confidence interval. In fact, you would not be very surprised to discover that Runner 2 was intrinsically slower than Runner 1. You do not have sufficient data to find a very narrow band for the difference in underlying ability – although, in practice, the selection of one athlete over another for a team often depends on evidence of this type.

A general method

Suppose that values of two random variables, X_1 and X_2, are measured on random samples of sizes n_1 and n_2. Let:

» X_1 and X_2 be distributed normally in the population with a common variance

» X_1 and X_2 be independent of each other in the population

» s^2 be the pooled-sample estimate of the common population variance of X_1 and X_2

» τ_a be the two-sided a% critical value for the t-distribution with $(n_1 + n_2 - 2)$ degrees of freedom.

Then a $(100 - a)$% confidence interval for the difference in the means μ_1 and μ_2 of X_1 and X_2 is given by

$$(\overline{x}_1 - \overline{x}_2) - \tau_a s\sqrt{\frac{1}{n_i} + \frac{1}{n_2}} < (\mu_1 - \mu_2) < (\overline{x}_1 - \overline{x}_2) + \tau_a s\sqrt{\frac{1}{n_1} + \frac{1}{n_2}}.$$

Exercise 2B

In questions **1–3**, you are expected to make a sensible choice of significance level for the hypothesis tests involved. Remember that the 5% level is conventional in scientific contexts.

(M) **1** A species of finch has subspecies on two different Galapagos Islands.

The weights of a sample of finches from each island are listed below.

Weights from Daphne Major (grams)											
64	67	64	61	68	61	61	67	70	62	66	63
Weights from Daphne Minor (grams)											
63	61	65	65	61	63	63					

(i) Is there evidence that the finches on Daphne Major are heavier on average than those on Daphne Minor?

(ii) What assumptions do you need to make? Are they reasonable?

(M) **2** Two groups of subjects are asked to volunteer for a psychology experiment. One group is told that they will be paid $1 for participating; the other that they will be paid $20. The experiment consists of a rather dull task that must be repeated for one hour: subjects are then asked to rate how interesting the task was, on a scale from 1 to 10, with 10 being the most interesting.

(i) Test, using a 2–sample *t*-test, the hypothesis that the task was found more interesting by those who were paid less.

The ratings of the two groups were:

Paid $1	4	2	6	3	5	1	6	4
	3	5	5	3	2	2	6	
	1	7	2	4	5	4	5	
Paid $20	3	5	2	1	3	4	5	
	2	1	1	3	4	4	4	
	5	2	3	2	1	3		

(ii) State and comment on the assumptions you are making in order to carry out this test.

(iii) Could you devise a paired design for this experiment?

3 A college careers officer investigates the income at age 24 of a group of students who left school at 16, and a group who stayed on to take A Levels.

The results he finds (measuring incomes in $ per week) are summarised in this table.

	16-year-old leavers	A Level leavers
Mean	156	164
Sample estimate of variance	673	593
Sample size	37	28

(i) Show, using a 2-sample t-test, that the hypothesis that those staying on at school have higher incomes at age 24 is rejected, on evidence of this sample. What assumptions are you making for a 2-sample t-test to be appropriate? How plausible are they?

(ii) What other difference between the two groups inevitably exists that might explain this unexpected result? How could you design an experiment to eliminate this effect?

In questions **4–6**, you need to decide whether the data are from an experiment with a paired or an unpaired design. You are expected to make a sensible choice of significance level for the hypothesis tests involved. Remember that the 5% level is conventional in scientific contexts.

4 Amongst all praying mantises, females are on average 7 centimetres longer than males. A new variety of mantis has been bred, the insects of which are supposed to be more nearly equal in size.

Test the hypothesis that the difference between male and female average lengths is less than 7 centimetres, using the lengths in centimetres of the sample of twelve males and twelve females shown below. State clearly the assumptions you are making in your test.

Males	18.4	15.1	11.9	16.3	8.7	13.0
	10.1	20.2	14.2	6.2	16.9	8.8
Females	28.2	24.2	12.4	23.4	13.2	21.4
	10.3	23.5	16.8	11.9	15.5	13.2

Answers to exercises are available at www.hoddereducation.com/cambridgeextras

5 In fact, the data given in question **4** are paired: each male mantis is paired with its mate in the following way:

Pair	1	2	3	4	5	6
Male	18.4	15.1	11.9	16.3	8.7	13.0
Female	28.2	24.2	12.4	23.4	13.2	21.4
Pair	7	8	9	10	11	12
Male	10.1	20.2	14.2	6.2	16.9	8.8
Female	10.3	23.5	16.8	11.9	15.5	13.2

(i) Test the hypothesis that male mantises are on average less than 7 centimetres smaller than their mates.

(ii) Explain clearly what assumptions you make in this case, and how these assumptions differ from those you made in question **4**.

(iii) Why are the results you obtain different in this case from those you found in question **4**?

6 It is known from many studies that the best current post–operative treatment reduces stays in hospital after major operations, compared with untreated patients, by an average of 6.2 days. A new treatment is proposed, with the hypothesis that this new treatment will reduce stays in hospital by more than 6.2 days on average, and a trial is conducted on two groups of patients who have just undergone major operations. The results are shown below.

Days stayed in hospital by untreated patients						
35	33	27	27	25	31	27
36	46	32	27	16	28	
Days stayed in hospital by patients given new treatment						
18	18	23	24	22	14	28
29	24	16	18			

Test the hypothesis given, clearly stating the assumptions you are making.

In questions **7–9**, you need to decide whether the data are from an experiment with a paired or an unpaired design.

7 The masses, in grams, of nine hens' eggs and eight ducks' eggs are recorded below.

Hens	42	47	45	41	48	39	46	45	48
Ducks	45	47	51	46	49	53	53	48	

(i) Construct a 95% confidence interval for the difference in mean masses of hens' eggs and ducks' eggs.

(ii) State the assumptions you are making in constructing this confidence interval.

8 A group of rowers and a group of chess players have their resting pulse rates measured. These data are shown below.

Rowers	65	73	71	80	61	77	83	70	76
Chess players	117	93	68	92	73	102	85		

Construct a one-sided 95% confidence interval, giving an upper limit for the extent to which the mean resting pulse rate of chess players exceeds that of rowers.

9 The amount, p, of infestation of maize fields by root nematodes, in grams of the pest per square metre, is measured in randomly chosen square metre areas on 33 maize farms. Some of the farms have sprayed the crops with a new pesticide. The measurements are summarised in the table below.

	Using new pesticides	Not using new pesticides
Number of farms	14	19
Σp	8831	16573
Σp^2	5692287	14908662

Construct a 90% confidence interval for the difference in mean infestation between the sprayed and unsprayed crops.

10 Fish of a certain species live in two separate lakes, A and B. A zoologist claims that the mean length of fish in A is greater than the mean length of fish in B. To test his claim, he catches a random sample of 8 fish from A and a random sample of 6 fish from B. The lengths of the 8 fish from A, in appropriate units, are as follows.

15.3   12.0   15.1   11.2   14.4   13.8   12.4   11.8

Assuming a normal distribution, find a 95% confidence interval for the mean length of fish in A.

The lengths of the 6 fish from B, in the same units, are as follows.

15.0   10.7   13.6   12.4   11.6   12.6

Stating any assumptions that you make, test at the 5% significance level whether the mean length of fish in A is greater than the mean length of fish in B.

Calculate a 95% confidence interval for the difference in the mean lengths of fish from A and from B.

Cambridge International AS & A Level Further Mathematics
9231 Paper 22 Q11 November 2014

Answers to exercises are available at www.hoddereducation.com/cambridgeextras

2.8 Using the normal distribution with two samples

In studying 2-sample *t*-tests, you had to make the assumption that the variance in the population of the random variable you were sampling was the same in both conditions. You then estimated this common variance from the two samples. However, there are some situations in which you know the variance of the whole population and you can use this information in a hypothesis test or in constructing confidence intervals.

For instance, it may be that, before the ability of the maths class to estimate one minute was tested (see page 43), extensive tests were conducted that determined that, in the school population as a whole, students' estimated minutes are normally distributed and have a standard deviation of 7.42 seconds. You are testing the hypotheses:

H_0: There is no difference between the mean of people's estimates of one minute before and after lunch.

H_1: After lunch, the mean of people's estimates of one minute tends to be shorter than before lunch.

But you can now make the assumption that people's estimates of one minute are normally distributed with standard deviation 7.42.

The null hypothesis implies that before-lunch and after-lunch estimates have distributions $N(\mu, 7.42^2)$ where μ is the common mean asserted by the null hypothesis. With this assumption, the mean of the 24 before-lunch estimates has distribution

$$\bar{X} \sim N\left(\mu, \frac{7.42^2}{24}\right)$$

and the mean of the 22 after-lunch estimates has distribution

$$\bar{Y} \sim N\left(\mu, \frac{7.42^2}{22}\right).$$

The distribution of the difference of the two sample means is therefore

$$\bar{X} - \bar{Y} \sim N\left(0, \frac{7.42^2}{24} + \frac{7.42^2}{22}\right).$$

In *Probability & Statistics 2* you considered hypothesis tests with the normal distribution: if X has distribution $N(0, \sigma^2)$, where variance σ^2 is known, then the test statistic $\frac{X}{\sigma}$ has the standard normal distribution, $N(0, 1)$. The test statistic here is

$$\frac{\bar{X} - \bar{Y}}{\sqrt{\frac{7.42^2}{24} + \frac{7.42^2}{22}}} = \frac{\bar{X} - \bar{Y}}{7.42\sqrt{\frac{1}{24} + \frac{1}{22}}}.$$

With the data used in the example on page 43, $\bar{x} = 51.542$ and $\bar{y} = 47.909$, so the test statistic has the value

$$\frac{51.542 - 47.909}{7.42\sqrt{\frac{1}{24} + \frac{1}{22}}} = 1.659.$$

The critical region for a one-tailed test at the 5% significance level for the standard normal distribution is $z > 1.645$. In this case, since $1.659 > 1.645$, you reject the null hypothesis and accept the alternative, that the after-lunch times are shorter than the before-lunch times.

Different known variances for the two samples

Alternatively, you might know separately the variances of the populations from which each sample was drawn, where these need not be the same.

Suppose there are two machines in a factory. The first is a high-accuracy machine, which produces bolts with radii that are normally distributed with standard deviation 0.052 mm. The second is a lower-accuracy machine, producing washers with internal radii that are normally distributed with standard deviation 0.172 mm. Both machines are adjustable to produce components with different radii, but today they are supposed to be set so that the high-accuracy machine produces bolts with radii 2 mm smaller than the internal radii of the washers produced by the low-accuracy machine.

To check whether the setting is correct, a sample of components is taken from each machine, and the radius of each measured. The results are shown in the following table.

Radii of bolts from high-accuracy machine (mm)								
8.42	8.21	8.29	8.31	8.25	8.38	8.29		
Internal radii of washers from low-accuracy machine (mm)								
10.32	10.12	9.98	10.09	10.57	10.49	10.10	10.28	10.35

You are testing the hypotheses:

H_0: The mean radius of the bolts being produced is 2 mm less than the mean internal radius of the washers being produced.

H_1: The mean radius of the bolts and the mean internal radius of the washers being produced do not differ by 2 mm.

You can assume that the radii of the components being produced by each machine are normally distributed with the standard deviations given above.

If X_W denotes the internal radius of the washer, and X_B the radius of a bolt, what is the distribution of the sample statistic $X_W - X_B$?

Answers to exercises are available at www.hoddereducation.com/cambridgeextras

With the assumptions stated, the mean internal radius of nine washers from the low-accuracy machine has distribution

$$\overline{X}_W \sim N\left(\mu_W, \frac{0.172^2}{9}\right)$$

where μ_W is the mean internal radius of the washers.

Similarly, the mean radius of seven bolts from the high-accuracy machine has distribution

$$\overline{X}_B \sim N\left(\mu_B, \frac{0.052^2}{7}\right)$$

where μ_B is the mean radius of the bolts.

The distribution of the difference of the two sample means is therefore

$$\overline{X}_W - \overline{X}_B \sim N\left(\mu_W - \mu_B, \frac{0.172^2}{9} + \frac{0.052^2}{7}\right).$$

The null hypothesis then states that $\mu_W - \mu_B = 2$ and so, if the null hypothesis is true

$$\overline{X}_W - \overline{X}_B \sim N\left(2, \frac{0.172^2}{9} + \frac{0.052^2}{7}\right).$$

Therefore, the test statistic

$$\frac{\overline{X}_W - \overline{X}_B - 2}{\sqrt{\frac{0.172^2}{9} + \frac{0.052^2}{7}}}$$

has a standard normal distribution.

In this case $\overline{X}_W = 10.256$ and $\overline{X}_B = 8.307$, so the test statistic is

$$\frac{10.256 - 8.307 - 2}{\sqrt{\frac{0.172^2}{9} + \frac{0.052^2}{7}}} = -0.84.$$

The critical region for a two-tailed test with the standard normal distribution at the 5% significance level is $z > 1.96$ or $z < -1.96$, and so here, since $-1.96 < -0.84 < 1.96$, you will accept the null hypothesis that the machines are correctly set.

You can use the same data to construct a 95% confidence interval for the difference in mean radii being produced by the two machines. You saw above that the distribution of the difference of the two sample means is

$$\overline{X}_W - \overline{X}_B \sim N\left(\mu_W - \mu_B, \frac{0.172^2}{9} + \frac{0.052^2}{7}\right)$$

so that

$$P\left(-1.96 < \frac{(\bar{X}_W - \bar{X}_B) - (\mu_W - \mu_B)}{\sqrt{\frac{0.172^2}{9} + \frac{0.052^2}{7}}} < 1.96\right) = 0.95$$

since 1.96 is the two-tailed 5% critical value for the standard normal distribution.

The confidence interval is therefore

$$(\bar{x}_W - \bar{x}_B) - 1.96\sqrt{\frac{0.172^2}{9} + \frac{0.052^2}{7}} < (\mu_W - \mu_B)$$

$$< (\bar{x}_W - \bar{x}_B) + 1.96\sqrt{\frac{0.172^2}{9} + \frac{0.052^2}{7}}.$$

With the values $\bar{x}_W = 10.256$ *and* $\bar{x}_B = 8.307,$ this is

$$1.83 < \mu_W - \mu_B < 2.07.$$

Summary: known variances

The null hypothesis is:

H_0: The difference between the means of the random variable in the two conditions is $(\mu_1 - \mu_2)$.

If

» the random variable, X, has a normal distribution in each condition
» $\bar{X}_1$ and $\bar{X}_2$ are the means of the samples in the two conditions
» n_1 and n_2 are the sizes of these samples
» σ_1 and σ_2 are the known standard deviations of the random variables X_1 and $X_2,$

then the test statistic is

$$\frac{\bar{X}_1 - \bar{X}_2 - (\mu_1 - \mu_2)}{\sqrt{\frac{\sigma_1^2}{n_1} + \frac{\sigma_2^2}{n_2}}}$$

which has a standard normal distribution $N(0, 1)$.

In the same situation, a $(100 - a)\%$ confidence interval for the difference of the means in the two conditions is

$$(\bar{x}_1 - \bar{x}_2) - z_a\sqrt{\frac{\sigma_1^2}{n_1} + \frac{\sigma_2^2}{n_2}} < \mu_1 - \mu_2 < (\bar{x}_1 - \bar{x}_2) + z_a\sqrt{\frac{\sigma_1^2}{n_1} + \frac{\sigma_2^2}{n_2}}$$

where z_a is the two-sided $a\%$ critical value for the standard normal distribution.

Answers to exercises are available at www.hoddereducation.com/cambridgeextras

If the variances in the two conditions are known and equal, each distribution having standard deviation σ, then the test statistic simplifies to

$$\frac{(\bar{X}_1 - \bar{X}_2) - (\mu_1 - \mu_2)}{\sigma\sqrt{\dfrac{1}{n_1} + \dfrac{1}{n_2}}}$$

and the confidence interval to

$$(\bar{x}_1 - \bar{x}_2) - z_a\sigma\sqrt{\frac{1}{n_1} + \frac{1}{n_2}} < \mu_1 - \mu_2 < (\bar{x}_1 - \bar{x}_2) + z_a\sigma\sqrt{\frac{1}{n_1} + \frac{1}{n_2}}.$$

Remember you can do the same sort of thing for the paired test, with null hypothesis:

H_0: The difference between the values of the random variable in the two conditions has mean d.

If

» the difference, X, between the pairs of values of the random variable in the two conditions has a normal distribution

» $\bar{X}$ is the mean of X

» σ is the known standard deviation of X

» n is the size of the sample

then the test statistic is $\dfrac{\bar{X} - d}{\dfrac{\sigma}{\sqrt{n}}}$, which has a standard normal distribution, N(0, 1).

In the same situation, a $(100 - a)\%$ confidence interval for the mean difference between the two conditions is

$$\bar{X} - z_a\frac{\sigma}{\sqrt{n}} < \mu_1 - \mu_2 < \bar{X} + z_a\frac{\sigma}{\sqrt{n}}$$

where z_a is the two-sided $a\%$ critical value for the standard normal distribution.

Note

This situation – where you know the variance of the difference between random variables in the two conditions – is very unlikely to arise in practice. Knowing the variance of the random variable in each condition separately is not sufficient, as the point of pairing is that the variable is not independently measured in the two conditions. Thus in a paired test, if X_1 and X_2 represent the random variable in the two conditions, $\mathrm{Var}[X_1 - X_2] \neq \mathrm{Var}[X_1] + \mathrm{Var}[X_2]$.

It may have occurred to you that, in fact, the examples you have considered so far in this chapter were rather contrived, and that is not surprising: it is difficult to think of circumstances where the tests described here actually apply – why should you know the variances of the distributions, but not their means? This does not mean that you have been wasting your time, however: the importance of this technique is seen in the next section.

2.9 Tests with large samples

When you used the *t*-distribution for testing hypotheses about differences between means and for constructing confidence intervals for the difference between means, you had to make the assumptions

» the underlying variables are normally distributed

» the variables have a common variance.

In many situations where you want to test hypotheses or construct confidence intervals, these assumptions do not hold. Fortunately, provided that a large sample is available, these assumptions are not essential.

1 The central limit theorem says that, for large sample sizes, even when a variable does not have a normal distribution, its sample mean is approximately normally distributed.

2 In general, the larger the sample being used, the smaller the error made in assuming that the population has exactly the variance given by its sample estimate.

The tests discussed so far in this chapter, which assume normally distributed variables with known variances, will therefore all give sensible results for large samples, even where the underlying variables are not normally distributed and the variances are not known, but must be estimated from the samples. As a rule of thumb, sample sizes of 30 or so are large enough for this approximation to be reasonable – although this will obviously depend on how non-normal the underlying distributions are. In many situations the approximation will be justified with substantially smaller samples.

> ## Note
>
> Even where it cannot be assumed that the underlying variables have a common variance, but where large samples are available, separate population variances can be estimated and used as if they were the actual population variances.

Example 2.5

Smartos are small sweets. Smartos tubes are filled by two different machines. A sample of tubes filled by each machine and the Smartos in each tube are counted. The results are given in the table below.

Number of Smartos per tube	39	40	41	42	43	Total
Frequency in machine A sample	17	23	35	31	27	**133**
Frequency in machine B sample	21	18	41	39	19	**138**

Assuming a common variance for the numbers of Smartos per tube produced by the machines, construct a 98% confidence interval for the difference in the mean number of Smartos in the tubes produced by the two machines.

▲ Figure 2.8

➜

Answers to exercises are available at www.hoddereducation.com/cambridgeextras

Solution

Here, you certainly cannot assume that the distribution of the number of Smartos per tube is normal since it is a discrete variable. Nonetheless, provided the sample size is large, and you are told that there is a common variance in the two conditions, you can assume that the statistic

$$\frac{(\bar{X}_A - \bar{X}_B) - (\mu_A - \mu_B)}{S\sqrt{\dfrac{1}{n_A} + \dfrac{1}{n_B}}}$$

has approximately a standard normal distribution. ($\bar{X}_A$ and $\bar{X}_B$ are the means of the numbers of Smartos per tube in the sample from each machine, S^2 is the pooled-sample estimator of the common variance, n_A and n_B are the sizes of each sample and μ_A and μ_B are the true mean numbers of Smartos produced by the two machines.)

The two-tailed 2% critical value for the standard normal distribution is 2.326, so that for approximately 98% of samples

$$-2.326 < \frac{(\bar{x}_A - \bar{x}_B) - (\mu_A - \mu_B)}{s\sqrt{\dfrac{1}{n_A} + \dfrac{1}{n_B}}} < 2.326.$$

That is, an approximate 98% confidence interval for the difference between the true mean numbers is

$$(\bar{x}_A - \bar{x}_B) - 2.326s\sqrt{\frac{1}{n_A} + \frac{1}{n_B}} < \mu_A - \mu_B < (\bar{x}_A - \bar{x}_B) + 2.326s\sqrt{\frac{1}{n_A} + \frac{1}{n_B}}.$$

Here

$$n_A = 133,\ \bar{x}_A = 41.2105,\ s_A = 1.3030$$

$$n_B = 138,\ \bar{x}_B = 41.1232,\ s_B = 1.2525$$

so that

$$s = \sqrt{\frac{132 \times 1.3030^2 + 137 \times 1.2525^2}{133 + 138 - 2}} = 1.2775$$

and so the confidence interval is

$$-0.2738 < \mu_A - \mu_B < 0.4484.$$

General procedure for large samples

Unpaired design

In an experiment with an unpaired design, a test of the hypothesis:

H_0: The difference between the means in the two conditions is δ
uses data from two samples, one taken in each condition. These data are summarised by

» $\bar{X}_1$ and $\bar{X}_2$, the means of the samples in the two conditions
» n_1 and n_2, the sizes of these samples
» S_1^2 and S_2^2, the sample estimators of the variances in the two conditions.

The test statistic is

$$\frac{\bar{X}_1 - \bar{X}_2 - \delta}{\sqrt{\dfrac{S_1^2}{n_1} + \dfrac{S_2^2}{n_2}}}$$

which has approximately a standard normal distribution, if the sample size is large.

In the same situation, a $(100 - a)\%$ confidence interval for the difference of the means in the two conditions is

$$(\bar{x}_1 - \bar{x}_2) - z_a \sqrt{\frac{s_1^2}{n_1} + \frac{s_2^2}{n_2}} < \mu_1 - \mu_2 < (\bar{x}_1 - \bar{x}_2) + z_a \sqrt{\frac{s_1^2}{n_1} + \frac{s_2^2}{n_2}}$$

where z_a is the two-sided $a\%$ critical value for the standard normal distribution.

Special cases

» If you know that the variances in the two conditions have the values of σ_1^2 and σ_2^2 these values should be used instead of the sample estimates s_1^2 and s_2^2 in the test statistic and the confidence limits.

» If you believe that the variances are the same in the two conditions you use the statistic

$$\frac{\bar{X}_1 - \bar{X}_2 - \delta}{S\sqrt{\dfrac{1}{n_1} + \dfrac{1}{n_2}}}$$

where $S^2 = \dfrac{(n_1 - 1)S_1^2 + (n_2 - 1)S_2^2}{(n_1 + n_2 - 2)}$ is the pooled sample variance estimate.

Similarly the confidence interval in this case is

$$(\bar{x}_1 - \bar{x}_2) - z_a s \sqrt{\frac{1}{n_1} + \frac{1}{n_2}} < \mu_1 - \mu_2 < (\bar{x}_1 - \bar{x}_2) + z_a s \sqrt{\frac{1}{n_1} + \frac{1}{n_2}}.$$

Answers to exercises are available at www.hoddereducation.com/cambridgeextras

Paired design

In an experiment with a paired design a test of the hypothesis:

H_0: The mean difference between variables in the two conditions is δ, uses data summarised by

» $\bar{D}$, the mean of the sample differences between the two conditions
» n, the size of the samples
» S^2, the sample estimator of the variance of the differences between the two conditions.

The test statistic is $\dfrac{\bar{D} - \delta}{\frac{S}{\sqrt{n}}}$, which has an approximately standard normal distribution if the sample size is large.

In the same situation, a $(100 - a)\%$ confidence interval for the difference of the means in the two conditions is

$$\bar{d} - z_a \frac{s}{\sqrt{n}} < \delta < \bar{d} + z_a \frac{s}{\sqrt{n}}$$

where z_a is the two-sided $a\%$ critical value for the standard normal distribution.

Special case

If you know that the variance of the difference between the two conditions has the value σ^2 this value should be used instead of the sample estimate s^2 in the test statistic and the confidence limits.

Exercise 2C

1 It has been suggested that, although the number of eggs laid by female hedgesparrows varies widely among birds, individual birds tend to lay more eggs each year that they breed. A sample of 162 female birds have their nests examined in two successive years, and the change in the number of eggs is recorded. The frequency of each change is recorded in the table below.

Change in number of eggs	−2	−1	0	1	2
Frequency	4	27	77	43	11

Use these data to construct a confidence interval for the mean increase in numbers of eggs per year, stating clearly the assumptions you are making.

2 A bank is deciding whether to introduce a new system for their cashiers. It is only worth the expense if it will reduce the average waiting time by a minute or more. They survey the waiting times for a sample of their customers in a branch and then, after introducing a trial run of the new system in the same branch, resurvey the waiting times. The results of the two surveys are shown in the table on the next page.

	Frequency	
Minutes in queue	With old system	With new system
0	33	49
1	57	58
2	44	44
3	35	24
4	17	12
5	13	2
6	10	4
7	6	1
8	5	0
9	5	0
10	1	1
11	4	1
12	3	0
13	3	0

Test the hypothesis that the waiting times have been reduced by at least one minute, on average, under the new system, stating the assumptions you are making.

3   Two very long shafts fitted to a turbine need to be very accurately the same length, but they cannot be moved to compare them, and are difficult to measure. The engineer whose job it is to check the lengths adopts the following procedure: ten separate measurements are made of each shaft, by a process of which the result is a normal variable with mean equal to the true length of the shaft and with standard deviation 3 hundredths of a millimetre.

The measurements made in one check are listed in the table below. (Lengths are given as hundredths of a millimetre from the nominal length of the shaft.)

First shaft	+6	+2	+7	−1	+4	−3	+3	+2	+7	+7
Second shaft	−1	+3	+3	+5	−2	−4	+4	+2	−3	+0

Test the hypothesis that the two shafts have the same length.

4   In an attempt to redesign a combustion chamber, it is necessary to find the difference between the maximum inside and outside temperatures of the casing. The combustion process is rather variable from one ignition to another; in fact, the variance of the maximum temperature, in °C, inside the chamber is 3940 and the variance of the temperature of the outer casing is 2710.

Construct a confidence interval for the difference between mean inside and outside temperatures using the data below from a series or experimental ignitions.

Nine ignitions: maximum inside temp (°C)								
6870	6940	7010	6960	6890	6940	6950	6920	6920

Six ignitions: maximum outside temp (°C)								
6710	6730	6680	6670	6680	6620			

5 A new machine for packing matches into boxes has been delivered. It is supposed to be more accurate; that is, the variation in the number of matches it inserts into each box is less, but the factory manager also wants to check that the new machine is not extravagantly putting a higher average number of matches into each box. She takes a sample of boxes produced by the old and new machines and counts the frequency with which each number of matches arises. The results are given in the table.

Number of matches	40	41	42	43	44
Frequency with old machine	96	42	17	9	2
Frequency with new machine	65	107	19	0	0

Test the hypothesis that the two machines have equal means, stating the assumptions you are using.

6 In an attempt to decide whether a new feeding regime increases the average weight of young salmon in a fish farm, a sample of 165 fish fed in the usual way is weighed. The weights, x_i, in grams, of these fish are summarised by the figures

$$\sum_{i=1}^{165} x_i = 11\,774 \qquad \sum_{i=1}^{165} x_i^2 = 872\,308.$$

A sample of 74 fish is bred with the new feeding regime and their weights, y_i, in grams, are summarised by

$$\sum_{i=1}^{74} y_i = 5491 \qquad \sum_{i=1}^{74} y_i^2 = 409\,272.$$

(i) Test the hypothesis that the average weight of the salmon is higher under the new regime

  (a) assuming that the variance of their weights is unchanged

  (b) estimating the variance separately in each condition.

(ii) Comment on your results.

7 In an experiment to determine whether presenting a list of words alphabetically or in random order causes them to be remembered more easily, two groups of subjects are given such lists of 30 words to study for one minute. After a distracting task, the subjects are then asked to recall as many words as possible in one minute. The numbers of words recalled in each condition are as follows.

Words recalled	14	15	16	17	18	19	20	21	22	23	24	25	26	27	28
Frequency: random list	0	0	0	1	10	12	34	78	55	22	29	11	2	1	2
Frequency: alphabetical list	3	2	23	57	17	21	19	33	48	26	19	4	8	0	0

Tip: The bimodal nature of the distribution with an alphabetical list is because subjects use two different strategies for recall in this case.

(i) Test the hypothesis that there is no difference on average in the number of words recalled in the two conditions. State the assumptions that you are making.

(ii) Suggest a paired design for testing this hypothesis.

In questions **8–14**, you are expected to choose the appropriate technique for yourself.

8 For reasons of economy, the manufacturers of an electrical appliance wished to make an adjustment to one of its components. Before finally deciding whether to do so, the effect of the adjustment on the resistance of the component was assessed.

(i) At one factory, the resistance of ten such components selected at random were measured both before and after the adjustment. The results were as follows.

Component	Resistance (ohms) before adjustment	Resistance (ohms) after adjustment
1	37.7	41.0
2	42.1	47.8
3	44.2	44.9
4	35.2	39.6
5	38.6	45.8
6	43.2	45.9
7	47.3	49.6
8	35.9	38.7
9	43.7	44.5
10	42.4	49.1

Test at the 5% significance level whether there is a difference in mean resistance due to the adjustment.

Answers to exercises are available at www.hoddereducation.com/cambridgeextras

(ii) At another factory, twenty of the components were selected at random and ten of these were chosen at random to receive the adjustment. The resistances of the components were measured, with the following results.

Unadjusted component's resistance (ohms), x	Adjusted components resistance (ohms), y
42.3	49.2
35.3	44.8
44.3	39.5
42.0	47.7
37.6	39.9
43.6	44.4
35.8	49.3
47.6	38.8
43.1	45.8
38.5	45.7

(For information: $\Sigma x = 410.1$, $\Sigma x^2 = 16\,963.85$, $\Sigma y = 445.1$, $\Sigma y^2 = 19\,948.65$)

Test at the 5% significance level whether there is a difference between the mean resistances in each group, stating carefully the assumption you make about the underlying variances.

(iii) Explain clearly which of the two analyses gives better information and why.

9 Investigations are being made of the time taken to bring water to the boil in a large urn in a cafeteria. It is known that this time varies somewhat and that the variations may be accounted for by taking the boiling time to be normally distributed.

(i) The boiling times in minutes on ten randomly chosen occasions were 20.2, 17.8, 23.6, 21.1, 19.4, 19.6, 20.9, 20.0, 18.9, 20.3. Find a two-sided symmetrical 95% confidence interval for the true mean boiling time.

(ii) A second urn is acquired, for the boiling time may again be taken as normally distributed but with a possibly different mean. A random sample of 20 boiling times for the second urn is found to have mean 19.33 minutes.

Information from the manufacturers states that the *true* standard deviation of boiling time is 0.9 minutes for both urns. Assuming this is indeed correct, examine at the 5% level of significance whether the true mean boiling times for the two urns differ.

10 Two personal computers are being compared with respect to their performances in running typical jobs. Eight typical jobs selected at random are run on each computer. The table shows the values of a composite unit of performance for each job on each computer.

Job	Computer A	Computer B
1	214	203
2	198	202
3	222	216
4	206	218
5	194	185
6	236	224
7	219	213
8	210	212

It is desired to examine whether the mean performance for typical jobs is the same for each computer.

(i) State formally the null and alternative hypotheses that are being tested.

(ii) State an appropriate assumption concerning the underlying normality.

(iii) Carry out the test, using a 1% level of significance.

(iv) Provide a symmetrical two-sided 95% confidence interval for the difference between the mean performance times.

11 The central business district of a town is served by two railway stations, A and B. Part of a study is to examine whether the mean daily number of passengers arriving at station A during the morning peak period is the same as the corresponding average at station B. Counts were taken at station A for a random sample of 8 working days and at station B for a separate random sample of 12 working days, with the following results for the numbers of passengers arriving during the morning peak period.

Station A	1486	1529	1512	1540	1506	1464
Station B	1475	1497	1460	1478	1520	1473
Station A	1495	1502				
Station B	1449	1480	1503	1462	1474	1486

(i) State formally the null and alternative hypotheses that are to be tested.

(ii) State an appropriate assumption concerning underlying normality.

(iii) State a further necessary assumption concerning the underlying distributions.

(iv) Carry out the test, using a 5% level of significance.

(v) Suppose that, in a test situation such as this, the *true* variances of the underlying distributions were known. Outline *briefly* how the test would be conducted.

Answers to exercises are available at www.hoddereducation.com/cambridgeextras

12 A liquid product is sold in containers. The containers are filled by a machine. The volumes of liquid (in millilitres) in a random sample of six containers were found to be

497.8   501.4   500.2   500.8   498.3   500.0.

After an overhaul of the machine, the volumes (in millilitres) in a random sample of 11 containers were found to be

501.1   499.6   500.3   500.9   498.7   502.1   500.4   499.7
501.0   500.1   499.3.

It is desired to examine whether the average volume of liquid delivered to a container by the machine is the same after the overhaul as it was before.

(i)   State the assumptions that are necessary for the use of the customary *t*-test.

(ii)   State formally the null and alternative hypotheses that are to be tested.

(iii) Carry out the *t*-test, using a 5% level of significance.

(iv)   Discuss briefly which of the assumptions in part (i) is the least likely to be valid in practice and why.

PS

13 A railway station has a telephone enquiry office. The length of time, in minutes, taken to deal with any caller's enquiry is independent of that for all other callers and is modelled by the continuous random variable X with probability density function

$$f(x) = \frac{1}{4}xe^{-x/2} \qquad 0 \leqslant x < \infty.$$

(i)   Show that the mean of X is 4.

Reminder: The limit of $x^m e^{-x}$ as $x \to \infty$ is zero.

(ii)   You are now given that the variance of X is 8. State the mean and variance of T, the combined time for dealing with eight callers.

(iii) Explain why a normal random variable will provide a good approximation to the distribution of T.

(iv)   An attempt is made to improve the modelling. The detailed form of the X variable is discarded, although it is still believed appropriate to use a random variable whose variance is twice the mean. Denoting this mean by θ (and therefore the variance by 2θ), write down the parameters of the normal distribution that is now to be used as an approximation to the distribution of T. Deduce that, according to this distribution,

$$P\left(\frac{T - 8\theta}{4\sqrt{\theta}} < 1.645\right) = 0.95.$$

(v)   The combined time for dealing with eight callers is measured once and found to be 25 minutes. Show that, using the distribution in part (iv), the lower limit of a one-sided 95% confidence interval for θ is a solution of the equation

$$64\theta - 443.2964\theta + 625 = 0.$$

What does the other solution of this equation represent?

14 An inspector is examining the lengths of time taken to complete various routine tasks by employees who have been trained in two different ways. He wants to examine whether the two methods lead, overall, to the same times. Ten different tasks have been prepared. Each task is undertaken by a randomly selected employee who has been trained by method A and by a randomly selected employee who has been trained by method B. The times to completion, in minutes, are shown in the table.

Task	Time taken A employee	Time taken B employee
1	33.4	27.1
2	41.0	42.0
3	26.8	23.0
4	37.2	33.9
5	47.4	38.1
6	27.5	27.7
7	34.0	32.7
8	28.4	23.2
9	35.0	35.0
10	20.7	22.7

(i) Explain why these data should be analysed by a paired sample test.

(ii) What underlying distributional assumption is necessary for the paired sample t–test to be appropriate? Carry out this test at the 5% level of significance for the data above. Provide a two-sided 90% confidence interval for the true mean difference between times.

KEY POINTS

1 When the population standard deviation, σ, is not known and is estimated as being the sample standard deviation, s, and the distribution is normal, confidence intervals for μ are found using the t-distribution.

2 Two-sided confidence intervals for μ based on the t-distribution are given by

$$\bar{x} - k\frac{s}{\sqrt{n}} \text{ to } \bar{x} + k\frac{s}{\sqrt{n}}$$

3 The value of k for any confidence level can be found using t-distribution tables.

4 The value of s can be found using the formula

$$s^2 = \frac{S_{xx}}{n-1} = \sum_i \frac{(x_i - \bar{x})^2}{(n-1)}$$

Answers to exercises are available at www.hoddereducation.com/cambridgeextras

5   As with confidence intervals based on the normal distribution, confidence intervals for paired samples are formed in the same way, but the variable is now the difference between the paired values.

6   When the population standard deviation, σ, is not known and is estimated as being the sample standard deviation, s, and the distribution is normal, you can carry out a hypothesis test using a *t*-distribution.

7   The null hypothesis is the same as for the normal distribution
$H_0: \mu = \mu_0.$

8   You estimate the population standard deviation σ using the sample standard deviation s.

9   To carry out a hypothesis test on a single or paired sample, the test statistic $t = \dfrac{\bar{x} - \mu_0}{\frac{s}{\sqrt{n}}}$ is used.

10   You should now know how to test the hypothesis that the mean value of some random variable differs in two conditions, in a number of contexts. The test statistics used in the different contexts are given in the following table.

Variances	Sample size	Underlying distribution	Paired test	Unpaired test: variances equal	Unpaired test: variances unequal
Known	Any	Normal	$\dfrac{\bar{D}}{\frac{\sigma}{\sqrt{n}}} \sim \mathrm{N}(0, 1)$	$\dfrac{\bar{X} - \bar{Y}}{\sigma\sqrt{\frac{1}{n_X} + \frac{1}{n_Y}}} \sim \mathrm{N}(0, 1)$	$\dfrac{\bar{X} - \bar{Y}}{\sqrt{\frac{\sigma_x^2}{n_X} + \frac{\sigma_Y^2}{n_Y}}} \sim \mathrm{N}(0, 1)$
Known	Large	Need not be normal	$\dfrac{\bar{D}}{\frac{\sigma}{\sqrt{n}}} \approx \mathrm{N}(0, 1)$	$\dfrac{\bar{X} - \bar{Y}}{\sigma\sqrt{\frac{1}{n_X} + \frac{1}{n_Y}}} \approx \mathrm{N}(0, 1)$	
Not known	Any	Normal	$\dfrac{\bar{D}}{\frac{S}{\sqrt{n}}} \sim t_{n-1}$	$\dfrac{\bar{X} - \bar{Y}}{S\sqrt{\frac{1}{n_X} + \frac{1}{n_Y}}} \sim t_{n_X + n_Y - 2}$	No test discussed here is appropriate
Not known	Large	Need not be normal	$\dfrac{\bar{D}}{\frac{S}{\sqrt{n}}} \approx \mathrm{N}(0, 1)$	$\dfrac{\bar{X} - \bar{Y}}{S\sqrt{\frac{1}{n_X} + \frac{1}{n_Y}}} \approx \mathrm{N}(0, 1)$	$\dfrac{\bar{X} - \bar{Y}}{\sqrt{\frac{S_X^2}{n_X} + \frac{S_Y^2}{n_Y}}} \approx \mathrm{N}(0, 1)$
Known or not	Small	Not normal	No test discussed here is appropriate		

11   For an unpaired *t* test, the sample variance is estimated by $S^2 = \dfrac{(n_1 - 1)S_1^2 + (n_2 - 1)S_2^2}{(n_1 + n_2 - 2)}$

12   To test the hypothesis that two random variables differ by the amount δ simply replace $\bar{D}$ by $(\bar{D} - \delta)$ or $(\bar{X} - \bar{Y})$ by $(\bar{X} - \bar{Y} - \delta)$ in the numerator of the appropriate statistic.

13   Confidence intervals can be constructed from the test statistic by observing that a confidence interval, in the appropriate situation, for the difference of the means in the two conditions, is given by the interval:

(numerator of statistic) $\pm$ (critical value) $\times$ (denominator of statistic).

14   Note carefully the sections of this table with no appropriate test. It is just as important to know when not to conduct a test as how to do so!

LEARNING OUTCOMES

Now you have finished this chapter, you should be able to
- formulate hypotheses and apply a hypothesis test concerning the population mean using a small sample drawn from a normal population of unknown variance, using a t-test
- calculate a pooled estimate of a population variance from two samples
- formulate hypotheses concerning the difference of population means, and apply, as appropriate
  - a 2-sample t-test
  - a paired sample t-test
  - a test using a normal distribution
- know when samples from two populations should be considered as paired
- know the meaning of the term *confidence interval* for a parameter and associated language
- determine a confidence interval for a population mean, based on a small sample from a normal population with unknown variance, using a t-distribution
- understand the factors that affect the width of a confidence interval
- determine a confidence interval for a difference of population means, using a t-distribution or a normal distribution, as appropriate.

Answers to exercises are available at www.hoddereducation.com/cambridgeextras

3 Chi-squared tests

> ... the fact that the criterion we happen to use has a fine ancestry of ... statistical theorems does not justify its use. Such justification must come from empirical evidence that it works.
>
> *W.A. Shewhart (1891–1967)*

What kind of films do you enjoy?

To help it decide when to show trailers for future programmes, the management of a cinema asks a sample of its customers to fill in a brief questionnaire saying which type of film they enjoy. It wants to know whether there is any relationship between people's enjoyment of horror films and action movies.

> › How do you think the management should select the sample of customers?

3.1 The chi-squared test for a contingency table

The management of the cinema takes 150 randomly selected questionnaires and records whether those patrons enjoyed or did not enjoy horror films and action movies.

Observed frequency f_o	Enjoyed horror films	Did not enjoy horror films
Enjoyed action movies	51	41
Did not enjoy action movies	15	43

Note

You will meet larger contingency tables later in this chapter.

This method of presenting data is called a 2×2 **contingency table**. It is used where two variables (here 'attitude to horror films' and 'attitude to action movies') have been measured on a sample, and each variable can take two different values ('enjoy' or 'not enjoy').

The values of the variables fall into one or other of two categories. You want to determine the extent to which the variables are **related**.

It is conventional, and useful, to add the row and column totals in a contingency table; these are called the **marginal totals** of the table.

Observed frequency f_o	Enjoyed horror films	Did not enjoy horror films	Total
Enjoyed action movies	51	41	92
Did not enjoy action movies	15	43	58
Total	66	84	150

A formal version of the cinema management's question is, 'Is enjoyment of horror films independent of enjoyment of action movies?'. You can use the sample data to investigate this question.

You can estimate the probability that a randomly chosen cinema-goer will enjoy horror films as follows. The number of cinema-goers in the sample who enjoyed horror films is $51 + 15 = 66$.

So the proportion of cinema-goers who enjoyed horror films is $\frac{66}{150}$.

In a similar way, you can estimate the probability that a randomly chosen cinema-goer will enjoy action movies. The number of cinema-goers in the sample who enjoyed action movies is $51 + 41 = 92$.

Notice how you use the marginal totals 66 and 92 that were calculated previously.

So the proportion of cinema-goers who enjoyed action movies is $\frac{92}{150}$.

If people enjoyed horror films and action movies independently with the probabilities you have just estimated, then you would expect to find, for instance:

Answers to exercises are available at www.hoddereducation.com/cambridgeextras

Number of people enjoying both types

$$= 150 \times P(\text{a random person enjoying both types})$$

$$= 150 \times P(\text{enjoying horror}) \times P(\text{enjoying action})$$

$$= 150 \times \frac{66}{150} \times \frac{92}{150}$$

$$= \frac{6072}{150}$$

$$= 40.48$$

In the same way, you can calculate the number of people you would expect to correspond to each cell in the table.

Expected frequency f_e	Enjoyed horror films	Did not enjoy horror films	Total
Enjoyed action movies	$150 \times \frac{66}{150} \times \frac{92}{150} = 40.48$	$150 \times \frac{84}{150} \times \frac{92}{150} = 51.52$	92
Did not enjoy action movies	$150 \times \frac{66}{150} \times \frac{58}{150} = 25.52$	$150 \times \frac{84}{150} \times \frac{58}{150} = 32.48$	58
Total	66	84	150

Note that it is an inevitable consequence of this calculation that these expected figures have the same marginal totals as the sample data.

You are now in a position to test the original hypotheses, which you can state formally as:

H_0: Enjoyment of the two types of film is independent.

H_1: Enjoyment of the two types of film is not independent.

The expected frequencies were calculated assuming the null hypothesis is true. You know the actual sample frequencies and the aim is to decide whether those from the sample are so different from those calculated theoretically that the null hypothesis should be rejected.

A statistic that measures how far apart a set of observed frequencies is from the set expected under the null hypothesis is the χ^2 (chi-squared) statistic. It is given by the formula

> The χ^2 test statistic is denoted by X^2.

$$X^2 = \sum \frac{(f_o - f_e)^2}{f_e} = \sum \frac{(\text{observed frequency} - \text{expected frequency})^2}{\text{expected frequency}}$$

You can use this here; the observed and expected frequencies are summarised below.

Observed frequency f_o	Enjoyed horror	Did not enjoy horror
Enjoyed action	51	41
Did not enjoy action	15	43

Expected frequency f_e	Enjoyed horror	Did not enjoy horror
Enjoyed action	40.48	51.52
Did not enjoy action	25.52	32.48

The χ^2 statistic is

$$X^2 = \Sigma \frac{(f_o - f_e)^2}{f_e} = \frac{(51 - 40.48)^2}{40.48} + \frac{(41 - 51.52)^2}{51.52}$$

$$+ \frac{(15 - 25.52)^2}{25.52} + \frac{(43 - 32.48)^2}{32.48}$$

$$= \frac{(10.52)^2}{40.48} + \frac{(-10.52)^2}{51.52} + \frac{(-10.52)^2}{25.52} + \frac{(10.52)^2}{32.48} = 12.626$$

> **Note that the four numerators in this calculation are equal. This is not by chance, it will always happen with a 2 × 2 table. It provides you with a useful check and short cut when you are working out X^2.**

Following the usual hypothesis–testing methodology, you want to know whether a value for this statistic at least as large as 12.626 is likely to occur by chance when the null hypothesis is true. The critical value at the 10% significance level for this test statistic is 2.706.

> **You will see how to find critical values for a χ^2 test later in this chapter.**

Since 12.626 > 2.706, you reject the null hypothesis, H_0, and conclude that people's enjoyment of the two types of film is not independent or that the enjoyment of the two is **associated**.

Figure 3.1 shows you the relevant χ^2 distribution for this example, the critical region and the test statistic.

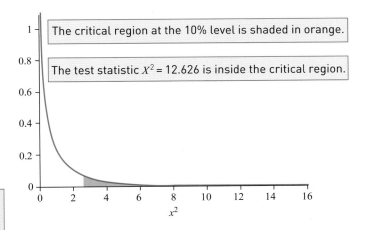

The critical region at the 10% level is shaded in orange.

The test statistic $X^2 = 12.626$ is inside the critical region.

▲ **Figure 3.1**

> **The information about the χ^2 distribution is for your interest – you do not need to use it to carry out the tests in this chapter.**
>
> **A standard normal variable is drawn from a normal population with mean 0 and variance 1.**

 Note

Notice that you cannot conclude that enjoying one type of film causes people to enjoy the other. The test is of whether enjoyment of the two types is associated. It could be that a third factor, such as bloodthirstiness, causes both, but you do not know. The test tells you nothing about causality.

The chi-squared distribution

The χ^2 distribution with n degrees of freedom is the distribution of the sum of the squares of n independent standard normal random variables.

You can use it to test how well a set of data matches a given distribution. A number of examples of such tests are covered in this chapter.

Answers to exercises are available at www.hoddereducation.com/cambridgeextras

These tests include that used in the example of the cinema-goers; that is, whether the two classifications used in a contingency table are independent of one another. The hypotheses for such a test are:

H_0: The two variables whose values are being measured are independent in the population.

H_1: The two variables whose values are being measured are not independent in the population.

In order to carry out this test, you need to know more about the χ^2 distribution.

Figure 3.1 is an example of a χ^2 distribution. The shape of the χ^2 distribution curve depends on the number of free variables involved, the degrees of freedom, v. To find the value for v in this case, you start off with the number of cells that must be filled and then subtract 1 degree of freedom for each restriction, derived from the data, that is placed on the frequencies. In the cinema example above, you are imposing the requirements that the total of the frequencies must be 150, and that the overall proportions of people enjoying horror films and action movies are $\frac{66}{150}$ and $\frac{92}{150}$, respectively.

Hence

$v = 4$ (number of cells)

$\quad - 1$ (total of frequencies is fixed by the data)

$\quad - 2$ (proportions of people enjoying each type are estimated from the data)

$\quad = 1.$

So Figure 3.1 shows the shape of the χ^2 distribution for 1 degree of freedom.

In general, for an $m \times n$ contingency table, the degrees of freedom are

$v = m \times n$ (number of cells)

$\quad - (m + n - 1)$ (row and column totals are fixed but row totals and column totals have the same sum.)

$\quad = mn - m - n + 1$

$\quad = (m - 1)(n - 1).$

As you will see later in the chapter, the calculation of the degrees of freedom varies from one χ^2 test to another.

Figure 3.2 shows the shape of the chi-squared distribution for $v = 1, 2, 3, 6,$ and 10 degrees of freedom.

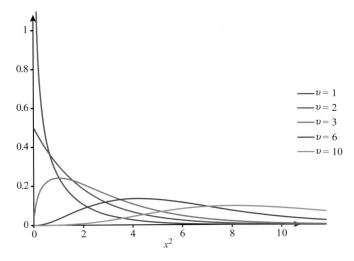

▲ **Figure 3.2**

You can see in Figure 3.3 a typical χ^2 distribution curve together with the critical region for a significance level of $1 - p$. An extract from a table of critical values of the χ^2 distribution for various degrees of freedom is also shown. The possible use of the left-hand tail probabilities ($p = 0.01$, $p = 0.025$, and so on) is discussed later in this chapter.

 Note

As you can see, the shape of the chi-squared distribution depends very much on the number of degrees of freedom. So the critical region also depends on the number of degrees of freedom.

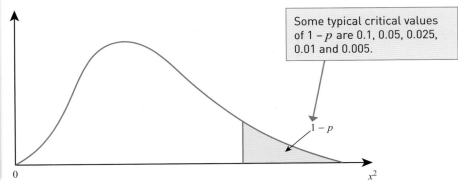

p	0.01	0.025	0.05	0.9	0.95	0.975	0.99	0.995	0.999
$v = 1$	0.0^31571	0.0^39821	0.0^23932	2.706	3.841	5.024	6.635	7.879	10.83
2	0.02010	0.05064	0.1026	4.605	5.991	7.378	9.210	10.60	13.82
3	0.1148	0.2158	0.3518	6.251	7.815	9.348	11.34	12.84	16.27
4	0.2971	0.4844	0.7107	7.779	9.488	11.14	13.28	14.86	18.47
5	0.5543	0.8312	1.145	9.236	11.07	12.83	15.09	16.75	20.51
6	0.8721	1.237	1.635	10.64	12.59	14.45	16.81	18.55	22.46
7	1.239	1.690	2.167	12.02	14.07	16.01	18.48	20.28	24.32
8	1.647	2.180	2.733	13.36	15.51	17.53	20.09	21.95	26.12
9	2.088	2.700	3.325	14.68	16.92	19.02	21.67	23.59	27.88

▲ **Figure 3.3** Critical values for the χ^2-distribution. **Tip:** 0.0^31571 means 0.0001571.

Answers to exercises are available at www.hoddereducation.com/cambridgeextras

Properties of the test statistic X^2

You have seen the test statistic is given by

$$X^2 = \sum_{All\ classes} \frac{(f_o - f_e)^2}{f_e}$$

Here are some points to notice:

» It is clear that as the difference between the expected values and the observed values increases then so will the value of this test statistic. Squaring the top gives due weight to any particularly large differences. It also means that all values are positive.

» Dividing $(f_e - f_o)^2$ by f_e has the effect of standardising that element, allowing for the fact that the larger the expected frequency within a class, the larger will be the difference between the observed and the expected.

» The usual convention in statistics is to use a Greek letter for a parent population parameter and the corresponding Roman letter for the equivalent sample statistic. Unfortunately, when it comes to χ^2, there is no Roman equivalent to the Greek letter χ since it translates into CH. Since X looks rather like χ, a sample statistic from a χ^2 population is denoted by X^2. (In the same way Christmas is sometimes abbreviated to χmas but written Xmas.)

For example:

Population parameters Greek letters	Sample statistics Roman letters
μ	m
σ	s
ρ	r

> ### Note
>
> An alternative notation that is often used is to call the expected frequency in the ith class Ei and the observed frequency in the ith class Oi.
>
> In this notation
>
> $$X^2 = \sum_i \frac{(O_i - E_i)^2}{E_i}$$

Continuing with tests on contingency tables

Example 3.1

The 4×3 contingency table on the next page shows the type of car (saloon, sports, hatchback or SUV) owned by 360 randomly chosen people, and the age category (under 30, 30–60, over 60) into which the owners fall.

Observed frequency f_o	Age of driver			Total
	under 30	30–60	over 60	
Saloon	10	67	57	134
Sports car	19	14	3	36
Hatchback	32	47	34	113
SUV	7	56	14	77
Total	68	184	108	360

(i) Write down appropriate hypotheses for a test to investigate whether type of car and owner's age are independent.

(ii) Calculate expected frequencies assuming that the null hypothesis is true.

(iii) Calculate the value of the test statistic X^2.

(iv) Find the critical value at the 5% significance level.

(v) Complete the test.

(vi) Comment on how the ownership of different types of car depends on the age of the owner.

Solution

(i) H_0: Car type is independent of owner's age.

H_1: Car type is not independent of owner's age.

(ii)

Expected frequency f_e	Age of driver			Total
	under 30	30–60	over 60	
Saloon	25.311	68.489	40.200	134
Sports car	6.800	18.400	10.800	36
Hatchback	21.344	57.756	33.900	113
SUV	14.544	39.356	23.100	77
Total	68	184	108	360

You need to calculate the expected frequencies in the table assuming that the null hypothesis is true.

Use the probability estimates given by the marginal totals.

For instance the expected frequency for hatchback and owner's age is over 60 is given by

$$360 \times \frac{113}{360} \times \frac{108}{360} = \frac{113 \times 108}{360} = 33.900$$

Note

This illustrates the general result for contingency tables

$$\text{expected frequency for a cell} = \frac{\text{product of marginal totals for that cell}}{\text{number of observations}}.$$

Answers to exercises are available at www.hoddereducation.com/cambridgeextras

Note

You need to check that all the frequencies are large enough to make the χ^2 distribution a good approximation to the distribution of the X^2 statistic. The usual rule of thumb is to require all the expected frequencies to be greater than 5.

This requirement is (just) satisfied here. However, you might be cautious in your conclusions if the X^2 statistic is very near the relevant critical value. If some of the cells have small expected frequencies, you should either collect more data or amalgamate some of the categories if it makes sense to do so. For instance, two adjacent age ranges could reasonably be combined, but two car types probably could not.

(iii) The value of the X^2 statistic is $X^2 = \sum \dfrac{(f_o - f_e)^2}{f_e}$.

The contributions of the various cells to this are shown in the table below.

An example of the calculation is

$$\frac{(10 - 25.311)^2}{25.311} = 9.262$$

for the top-left cell.

Contribution to test statistic	Age of driver		
	under 30	30–60	over 60
Saloon	9.262	0.032	7.021
Sports car	21.888	1.052	5.633
Hatchback	5.319	2.003	0.000
SUV	3.913	7.039	3.585

Total = 9.262 + 0.032 + 7.021 + 21.888 + ... + 3.585

$X^2 = 66.749$

The number of rows, m, is 4.

The number of columns, n, is 3.

(iv) The degrees of freedom are given by $v = (m - 1)(n - 1)$.

$v = (4 - 1) \times (3 - 1) = 6$

From the χ^2 tables, the critical value at the 5% level with 6 degrees of freedom is 12.59.

Note

You should always refer to the size of the contributions when commenting on the way that one variable is associated with the other (assuming, of course, that the conclusion to your test is that there is association).

(v) The observed X^2 statistic of 66.749 is greater than the critical value of 12.59. So the null hypothesis is rejected and the alternative hypothesis is accepted at the 5% significance level:

that car type is not independent of owner's age,

or that car type and owner's age are associated.

Note

You reject the null hypothesis if the test statistic is greater than the critical value.

(vi) In this case, the under-30 age group own fewer saloon cars and SUVs, more hatchbacks and many more sports cars than expected. Other cells with relatively large contributions to the X^2 statistic correspond to SUVs being owned more often than expected by 30–60-year-olds, and less often than expected by older or younger drivers, and over-60s owning more saloon cars and fewer sports cars than expected.

Technology note

Statistical software

You can use statistical software to carry out a χ^2 test for a contingency table.

In order for the software to process the test, you need to input the information in the table of observed frequencies. This consists of category names and the observed frequencies, so, in this case, it is the information in this table.

Observed frequency f_o	Age of driver		
	under 30	**30–60**	**over 60**
Saloon	10	67	57
Sports car	19	14	3
Hatchback	32	47	34
SUV	7	56	14

The software then carries out all the calculations. Here is a typical output.

Chi-squared test

	A	B	C	D
1		**under 30**	**30–60**	**over 60**
2	**Saloon**	25.3111	68.4889	40.2
3		9.2619	0.0324	7.0209
4		10	67	57
5	**Sports car**	6.8	18.4	10.8
6		21.8882	1.0522	5.633
7		19	14	3
8	**Hatchback**	21.3444	57.7556	33.9
9		5.3195	2.003	0.0003
10		32	47	34
11	**SUV**	14.5444	39.3556	23.1
12		3.9134	7.0394	3.5848
13		7	56	14

Result

Chi-squared test

df 6

X^2 .... 66.7493

p 0.0000

▲ **Figure 3.4**

Notice that the p-value is stated to be 0.0000. This requires some interpretation.

» The other output figures are given either to 4 decimal places or as whole numbers.

» So you can conclude that $p = 0.0000$ to 4 decimal places and therefore that $p < 0.00005$.

» So the result is significant even at the 0.01% significance level.

The output shown in Figure 3.4 includes the following information:
» the expected frequencies
» the contributions to the X^2 statistic
» the degrees of freedom
» the value of the X^2 statistic
» the critical value for the test, denoted by p; this is the equivalent of $1 - p$ in the tables.
❯ Identify where each piece of information is displayed.
❯ What other information is contained in the output box?

Using a spreadsheet

You can also use a spreadsheet to do the final stages of this test. To set it up, you would need to take the following steps:

1 Enter the same information as before; the variable categories and the observed frequencies.

2 Use a suitable formula to calculate the expected frequencies.

3 Combine classes as necessary if any expected frequencies are below 5.

4 Use a suitable formula to calculate the contributions to the test statistic.

5 Find the sum of the contributions.

6 Find the p-value using the formula provided with the spreadsheet, for example =CHISQ.DIST.RT(H1,6).

> Cell H1 contains the value of the test statistic, X^2.

> There are 6 degrees of freedom.

In this case, a typical spreadsheet gives the value of p as 1.894E−12; that is, 1.894×10^{-12}, so much less than the upper bound of 0.000 05 inferred from the statistical software.

Exercise 3A

1 A group of 330 students, some aged 13 and the rest aged 16, is asked 'What is your usual method of transport from home to school?'. The frequencies of each method of transport are shown in the table.

	Age 13	Age 16
Walk	43	35
Cycle	24	42
Bus	64	49
Car	41	32

(i) Find the total of each row and each column.

A student is going to carry out a test to determine whether method of transport is independent of age.

(ii) Show that the expected frequency for age 16 students who walk is 37.35.

(iii) Show that the expected frequency for age 13 students who cycle is 34.40.

(iv) Would you expect the method of transport to be independent of age?

2 A random sample of 80 students studying for a first aid exam was selected. The students were asked how many hours of revision they had done for the exam. The results are shown in the table, together with whether or not they passed the exam.

	Pass	Fail
Less than 10 hours	13	18
At least 10 hours	42	7

(i) Find the expected frequency for each cell of the table, for a test to determine whether hours of revision is independent of passing or failing.

(ii) Find the corresponding contributions to the chi-squared test statistic.

3 A group of 281 voters is asked to rate how good a job they think the President is doing. Each is also asked for the highest educational qualifications they have achieved. The frequencies with which responses occurred are shown in the table.

Rating of President	Highest qualifications achieved			
	None	IGCSE® or equivalent	A Level or equivalent	Degree or equivalent
Very poor	11	37	13	6
Poor	12	17	22	8
Moderate	7	11	25	10
Good	10	17	17	9
Very good	19	16	8	6

Use these figures to test whether there is an association between rating of the President and highest educational qualification achieved.

4 A medical insurance company office is the largest employer in a small town. When 37 randomly chosen people living in the town were asked where they worked and whether they belonged to the town's health club, 21 were found to work for the insurance company, of whom 15 also belonged to the health club, while 7 of the 16 not working for the insurance company belonged to the health club.

Test the hypothesis that health-club membership is independent of employment by the medical insurance company.

5 In a random sample of 163 adult males, 37 suffer from hay fever and 51 from asthma, both figures including 14 men who suffer from both. Test whether the two conditions are associated.

Answers to exercises are available at www.hoddereducation.com/cambridgeextras

3

6   A random sample of residents in a town took part in a survey. They were asked whether they would prefer the local council to spend money on improving the local bus service or on improving the quality of road surfaces. The responses are shown in the following table, classified according to the area of the town in which the residents live.

	Area 1	Area 2	Area 3
Local bus service	73	36	30
Road surfaces	47	44	20

Using a 5% significance level, test whether there is an association between the area lived in and preference for improving the local bus service or improving the quality of road surfaces.

Cambridge International AS & A Level Further Mathematics
9231 Paper 21 Q6 June 2011

M 7   A sample of 80 men and 150 women selected at random are tested for colour-blindness. Twelve of the men and five of the women are found to be colour-blind. Is there evidence at the 1% level that colour-blindness is related to gender?

M 8   Depressive illness is categorised as type I, II or III. In a group of depressive psychiatric patients, the length of time for which their symptoms are apparent is observed. The results are shown below.

Length of depressive episode	Type of symptoms		
	I	II	III
Brief	15	22	12
Average	30	19	26
Extended	7	13	21
Semi-permanent	6	9	11

Is the length of the depressive episode independent of the type of symptoms?

9   Customers were asked which of three brands of coffee, A, B and C, they prefer. For a random sample of 80 male customers and 60 female customers, the numbers preferring each brand are shown in the following table.

	A	B	C
Male	32	36	12
Female	18	30	12

Test, at the 5% significance level, whether there is a difference between coffee preferences of male and female customers.

A larger random sample is now taken. It consists of $80n$ male customers and $60n$ female customers, where n is a positive integer. It is found that the proportions choosing each brand are identical to those in the smaller sample. Find the least value of n that would lead to a different conclusion for the 5% significance level hypothesis test.

Cambridge International AS & A Level Further Mathematics
9231 Paper 21 Q10 November 2013

M 10 Public health officers are monitoring air quality over a large area. Air quality measurements using mobile instruments are made frequently by officers touring the area. The air quality is classified as poor, reasonable, good or excellent. The measurement sites are classified as being in residential areas, industrial areas, commercial areas or rural areas. The table shows a sample of frequencies over an extended period. The row and column totals and the grand total are also shown.

Measurement site	Air quality				
	Poor	Reasonable	Good	Excellent	Row totals
Residential	107	177	94	22	**400**
Industrial	87	128	74	19	**308**
Commercial	133	228	148	51	**560**
Rural	21	71	24	16	**132**
Column totals	**348**	**604**	**340**	**108**	**1400**

Examine at the 5% level of significance whether or not there is any association between measurement site and air quality, stating carefully the null and alternative hypotheses you are testing. Report briefly on your conclusions.

11 Random samples of employees are taken from two companies, A and B. Each employee is asked which of three types of coffee (Cappuccino, Latte, Ground) they prefer. The results are shown in the following table.

	Cappuccino	Latte	Ground
Company A	60	52	32
Company B	35	40	31

Test, at the 5% significance level, whether coffee preferences of employees are independent of their company.

Larger random samples, consisting of N times as many employees from each company, are taken. In each company, the proportions of employees preferring the three types of coffee remain unchanged. Find the least possible value of N that would lead to the conclusion, at the 1% significance level, that coffee preferences of employees are not independent of their company.

Cambridge International AS & A Level Further Mathematics
9231 Paper 21 Q10 June 2012

Answers to exercises are available at www.hoddereducation.com/cambridgeextras

12 The bank manager at a large branch was investigating the incidence of bad debts. Many loans had been made during the last year; the manager inspected the records of a random sample of 96 loans, and broadly classified them as satisfactory or unsatisfactory loans and as having been made to private individuals, small businesses or large businesses. The data were as follows.

	Satisfactory	Unsatisfactory
Private individual	22	5
Small business	34	11
Large business	21	3

(i) Discuss any problems that could occur in carrying out a χ^2 test to examine if there is any association between whether or not the loan was satisfactory and the type of customer to whom the loan was made.

(ii) State suitable null and alternative hypotheses for the test described in part (i).

(iii) Carry out a test at the 5% level of significance without combining any groups.

(iv) Explain which groups it might be best to combine and carry out the test again with these groups combined.

13 A survey of a random sample of 44 people is carried out. Their musical preferences are categorised as pop, classical or jazz. Their ages are categorised as under 20, 20 to 39, 40 to 59 and 60 or over. A test is to be carried out to examine whether there is any association between musical preference and age group. The results are as follows.

	Musical preference		
Age group	**Pop**	**Classical**	**Jazz**
Under 20	8	4	1
20–39	3	3	0
40–59	2	4	3
60 or over	1	7	8

(i) Calculate the expected frequencies for 'Under 20' and '60 or over' for pop music.

(ii) Explain why the test would not be valid using these four age categories.

(iii) State which categories it would be best to combine in order to carry out the test.

(iv) Using this combination, carry out the test at the 5% significance level.

(v) Discuss briefly how musical preferences vary between the combined age groups, as shown by the contributions to the test statistic.

14 A researcher is investigating the relationship between the political allegiance of university students and their childhood environment. He chooses a random sample of 100 students and finds that 60 have political allegiance to the Alliance party. He also classifies their childhood environment as rural or urban, and finds that 45 had a rural childhood. The researcher carries out a test, at the 10% significance level, on this data and finds that political allegiance is independent of childhood environment. Given that A is the number of students in the sample who both support the Alliance party and have a rural childhood, find the greatest and least possible values of A.

A second random sample of size $100N$, where N is an integer, is taken from the university student population. It is found that the proportions supporting the Alliance party from urban and rural childhoods are the same as in the first sample. Given that the value of A in the first sample was 29, find the greatest possible value of N that would lead to the same conclusion (that political allegiance is independent of childhood environment) from a test, at the 10% significance level, on this second set of data.

Cambridge International AS & A Level Further Mathematics
9231 Paper 23 Q11 June 2013

3.2 Goodness of fit tests

The χ^2 test is commonly used to see if a proposed model fits observed data.

> ❓
>
> ❯ Why is it that the model should fit the data and not the other way round?

Goodness of fit test for the uniform distribution

Are these dice biased?

Yong claims that the dice he is using are fair. His friend throws the dice a total of 120 times with these results:

Score	6	5	4	3	2	1
Frequency	12	16	15	23	24	30

Do these figures provide evidence that the dice are biased, or is this just the level of variation you would expect to occur naturally? Clearly, a formal statistical test is required.

The expected distribution of the results, based on the null hypothesis that the outcomes are not biased, is easily obtained. The probability of each outcome is $\frac{1}{6}$ and so the expectation for each number $120 \times \frac{1}{6} = 20$.

Outcome	1	2	3	4	5	6
Expected frequency, f_e	20	20	20	20	20	20

Answers to exercises are available at www.hoddereducation.com/cambridgeextras

You would not, however, expect exactly this result from 120 throws. Indeed, you would be very suspicious if somebody claimed to have obtained it, and might well disbelieve it. You expect random variation to produce small differences in the frequencies. The question is whether the quite large differences in the case of Yong's dice can be explained in this way or not.

When this is written in the formal language of statistical tests, it becomes:

Notice that these are the same hypotheses that you would use for a test on whether the dice are biased.

$H_0: p = \dfrac{1}{6}$ for each outcome.

$H_1: p \ne \dfrac{1}{6}$ for each outcome.

The expected frequencies are denoted by f_e and the observed frequencies by f_o. To measure how far the observed data are from the expected, you clearly need to consider the difference between the observed frequencies, f_o, and the expected, f_e. The measure that is used as a test statistic for this is denoted by X^2 and given by

$$X^2 = \sum_{\text{All classes}} \frac{(f_o - f_e)^2}{f_e}$$

You have already met this statistic when carrying out tests involving contingency tables earlier in this chapter. In this case, the calculation of X^2 is as follows.

Outcome	6	5	4	3	2	1
Observed frequency, f_o	12	16	15	23	24	30
Expected frequency, f_e	20	20	20	20	20	20
Difference, $f_o - f_e$	−8	−4	−5	3	4	10
$(f_o - f_e)^2$	64	16	25	9	16	100
$(f_o - f_e)^2/f_e$	3.2	0.8	1.25	0.45	0.8	5

$X^2 = 3.2 + 0.8 + 1.25 + 0.45 + 0.8 + 5 = 11.5$

The test statistic X^2 has the χ^2 (chi-squared) distribution. Critical values for this distribution are given in tables but, before you can use them, you have to think about two more points.

> What is to be the significance level of the test?

This should really have been set before any data were collected. Because it is unusual for dice to be biased, it would seem advisable to make the test rather strict, and so the 1% significance level is chosen.

> How many degrees of freedom are involved?

As mentioned in the section on contingency tables, the shape of the χ^2 distribution curve depends on the number of free variables involved, the degrees of freedom, ν. To find the value for ν you start off with the number of cells that must be filled and then subtract 1 degree of freedom for each restriction, derived from the data, that is placed on the frequencies.

In this case, there are six classes (corresponding to scores of 1, 2, 3, 4, 5 and 6) but since the total number of throws is fixed (120) the frequency in the last class can be worked out if you know those of the first five classes.

degrees of freedom

number of classes $\qquad$ v $\qquad = \qquad$ 6 $\qquad - \qquad$ 1

Looking in the tables for the 1% significance level and $v = 5$ gives a critical value of 15.09; see Figure 3.5.

number of restrictions

Since $11.5 < 15.09$, H_0 is accepted.

There is no reason at this significance level to believe that any number was any more likely to come up than any other. Yong's dice appear to be unbiased.

p	0.01	0.025	0.05	0.9	0.95	0.975	0.99	0.995	0.999
$v = 1$	0.0^31571	0.0^39821	0.0^23932	2.706	3.841	5.024	6.635	7.879	10.83
2	0.02010	0.05064	0.1026	4.605	5.991	7.378	9.210	10.60	13.82
3	0.1148	0.2158	0.3518	6.251	7.815	9.348	11.34	12.84	16.27
4	0.2971	0.4844	0.7107	7.779	9.488	11.14	13.28	14.86	18.47
5	0.5543	0.8312	1.145	9.236	11.07	12.83	15.09	16.75	20.51
6	0.8721	1.237	1.635	10.64	12.59	14.45	16.81	18.55	22.46
7	1.239	1.690	2.167	12.02	14.07	16.01	18.48	20.28	24.32
8	1.647	2.180	2.733	13.36	15.51	17.53	20.09	21.95	26.12
9	2.088	2.700	3.325	14.68	16.92	19.02	21.67	23.59	27.88

Note

This is a one-tailed test with only the right-hand tail under consideration. The interpretation of the left-hand tail (where the agreement seems to be too good) is discussed later in the chapter.

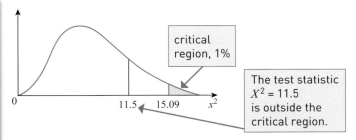

critical region, 1%

The test statistic $X^2 = 11.5$ is outside the critical region.

▲ Figure 3.5

Note

As with the test for a contingency table, the expected frequency of any class must be at least five. If a class has an expected frequency of less than five, then it must be grouped together with one or more other classes until their combined expected frequency is at least five.

When a particular distribution is fitted to the data, it may be necessary to estimate one or more parameters of the distribution. This, together with the restriction on the total, will reduce the number of degrees of freedom:

v = number of classes – number of estimated parameters – 1

Answers to exercises are available at www.hoddereducation.com/cambridgeextras

Goodness of fit test for the Poisson distribution

Example 3.2

The number of telephone calls made to a counselling service is thought to be modelled by the Poisson distribution. Data are collected on the number of calls received during one-hour periods, as shown in the table. Use these data to test at the 5% significance level whether a Poisson model is appropriate.

No. of calls per hour	0	1	2	3	4	5	6	Total
Frequency	6	13	26	14	7	4	0	70

Solution

H_0: The number of calls can be modelled by the Poisson distribution.

H_1: The number of calls cannot be modelled by the Poisson distribution.

Nothing is known about the form of the Poisson distribution, so the data must be used to estimate the Poisson parameter.

From the data, the mean number of calls per hour is

$$\frac{0 \times 6 + 1 \times 13 + 2 \times 26 + 3 \times 14 + 4 \times 7 + 5 \times 4}{70} = \frac{155}{70} = 2.214$$

The Poisson distribution with parameter $\lambda = 2.214$ is as follows.

x	$P(X = x)$	$70 \times P(X = x)$	Expected frequency
0	$e^{-2.214}$	70×0.1093	7.65
1	$e^{-2.214} \times \frac{2.214^1}{1!}$	70×0.2419	16.93
2	$e^{-2.214} \times \frac{2.214^2}{2!}$	70×0.2678	18.74
3	$e^{-2.214} \times \frac{2.214^3}{3!}$	70×0.1976	13.84
4	$e^{-2.214} \times \frac{2.214^4}{4!}$	70×0.1094	7.66
5	$e^{-2.214} \times \frac{2.214^5}{5!}$	70×0.0484	3.39
$\geqslant 6$	$1 - P(X < 6)$	70×0.0256	1.79

> ❗ The expected frequencies are not rounded to the nearest whole number. To do so would invalidate the test. Expected frequencies do not need to be integers.

The expected frequencies for the last two classes are both less than 5 but if they are put together to give an expected value of 5.2, the problem is overcome.

> ❗ The expected frequency for the last class is worked out as $1 - P(X < 6)$ and not as $P(X = 6)$, which would have cut off the right-hand tail of the distribution. The classes need to cover all *possible* outcomes, not just those that occurred in your survey.

The table for calculating the test statistic is shown below.

No. of calls, X	0	1	2	3	4	5+	Total
Observed frequency, f_o	6	13	26	14	7	4	70
Expected frequency, f_e	7.65	16.93	18.74	13.84	7.66	5.18	
$(f_o - f_e)$	−1.65	−3.93	7.26	0.16	−0.66	−1.18	
$(f_o - f_e)^2 / f_e$	0.3544	0.9126	2.8080	0.0020	0.0567	0.2702	4.440

> These values have been calculated from the exact expected frequencies, not the rounded figures given above.

$X^2 = 0.3544 + 0.9126 + 2.8080 + 0.0020 + 0.0567 + 0.2702 = 4.4040$

The degrees of freedom are

v = number of classes − number of estimated parameters − 1

No. of calls, X	0	1	2	3	4	5+	Total
Observed frequency, f_o	6	13	26	14	7	4	70

$$v \quad = \quad 6 \quad - \quad 1 \quad - \quad 1 \quad = \quad 4$$

> The number of classes is 6 because 2 of the original 7 classes have been combined.

> λ was estimated as 2.214, one restriction.

> The total frequency (70) is one restriction.

From the tables, the critical value for a significance level of 5% and 4 degrees of freedom is 9.488.

The calculated test statistic, $X^2 = 4.404$. Since $4.404 < 9.488$, H_0 is accepted.

The data are consistent with a Poisson distribution for the number of calls.

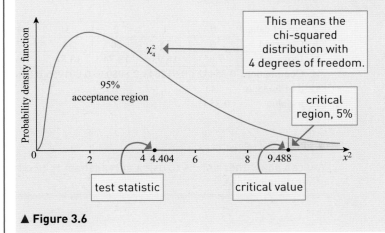

> This means the chi-squared distribution with 4 degrees of freedom.

Probability density function

χ_4^2

95% acceptance region

critical region, 5%

test statistic

critical value

▲ Figure 3.6

Answers to exercises are available at www.hoddereducation.com/cambridgeextras

Technology note

You can also use a spreadsheet to carry out the steps for a goodness of fit test for the Poisson model. To set it up, you would need to take the following steps.

1 Enter the variable categories and the observed frequencies. In this example they are

No. of calls, X	0	1	2	3	4	5	6
Observed frequency, f_o	6	13	26	14	7	4	0

In this case, the mean works out to be 2.21429.

2 Find the mean value.

3 Calculate the Poisson probabilities and use them to work out the expected frequencies.

This includes the final class, which is open-ended.

4 Combine classes as necessary if any expected frequencies are below 5.

5 Calculate the individual contributions to the test statistic.

6 Find the sum of the contributions.

7 Find the degrees of freedom.

8 Find the p-value using the formula provided with the spreadsheet, for example = CHISQ.DIST.RT.

A typical display for this example is shown below.

	A	B	C	D	E	F	G	H
1	No. of calls.	0	1	2	3	4	5	>=6
2	Observed	6	13	26	14	7	4	0
3	Expected	7.646	16.931	18.745	13.836	7.659	3.392	1.792
4								
5	Mean	2.21429						
6								
7	No. of calls.	0	1	2	3	4	>=5	Sum
8	Observed	6	13	26	14	7	4	70
9	Expected	7.65	16.93	18.74	13.84	7.66	5.18	70.00
10	Contribution	0.354423	0.912643	2.808036	0.001955	0.056694	0.270227	4.403979
11	d of freedom	4						
12	p-value	0.3541						
13								

▲ Figure 3.7

You can adapt these steps to use a spreadsheet to test for goodness of fit of other distributions.

Goodness of fit test for the binomial distribution

Example 3.3

An egg packaging firm has introduced a new box for its eggs. Each box holds six eggs. Unfortunately, it finds that the new box tends to mark the eggs. Data on the number of eggs marked in 100 boxes are collected.

No. of marked eggs	0	1	2	3	4	5	6	Total
No. of boxes, f_o	3	3	27	29	10	7	21	**100**

It is thought that the distribution may be modelled by the binomial distribution. Carry out a test on the data at the 0.5% significance level to determine whether the data can be modelled by the binomial distribution.

Solution

H_0: The number of marked eggs can be modelled by the binomial distribution.

H_1: The number of marked eggs cannot be modelled by the binomial distribution.

The binomial distribution has two parameters, n and p. The parameter n is clearly 6, but p is not known and so must be estimated from the data.

From the data, the mean number of marked eggs per box is

$$\frac{0 \times 3 + 1 \times 3 + 2 \times 27 + 3 \times 29 + 4 \times 10 + 5 \times 7 + 6 \times 21}{100} = 3.45$$

Since the population mean is np you may estimate p by putting $6p = 3.45$ estimated $p = 0.575$ and estimated $q = 1 - p = 0.425$.

These parameters are now used to calculate the expected frequencies of 0, 1, 2, ..., 6 marked eggs per box in 100 boxes.

x	$P(X = x)$		Expected frequency, f_e $100 \times P(X = x)$
0	0.425^6	0.0059	0.59
1	$6 \times 0.575^1 \times 0.425^5$	0.0478	4.78
2	$15 \times 0.575^2 \times 0.425^4$	0.1618	16.18
3	$20 \times 0.575^3 \times 0.425^3$	0.2919	29.19
4	$15 \times 0.575^4 \times 0.425^2$	0.2962	29.62
5	$6 \times 0.575^5 \times 0.425^1$	0.1603	16.03
6	0.575^6	0.0361	3.61

In this case, there are three classes with an expected frequency of less than 5. The class for $x = 0$ is combined with the class for $x = 1$, bringing the expected frequency just over 5, and the class for $x = 6$ is combined with that for $x = 5$.

➜

Answers to exercises are available at www.hoddereducation.com/cambridgeextras

No. of marked eggs, x	0, 1	2	3	4	5, 6	Total
Observed frequency, f_o	6	27	29	10	28	100
Expected frequency, f_e	5.37	16.18	29.19	29.62	19.64	100
$(f_o - f_e)$		0.63	10.82	−0.19	−19.62	8.36	
$(f_o - f_e)^2/f_e$		0.07	7.24	0.00	13.00	3.56	

The test statistic, $X^2 = 0.07 + 7.24 + 0.00 + 13.00 + 3.56 = 23.87$

The degrees of freedom

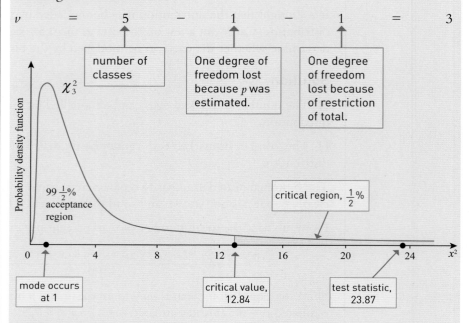

$$v \quad = \quad 5 \quad - \quad 1 \quad - \quad 1 \quad = \quad 3$$

number of classes

One degree of freedom lost because p was estimated.

One degree of freedom lost because of restriction of total.

χ_3^2

$99\frac{1}{2}\%$ acceptance region

critical region, $\frac{1}{2}\%$

mode occurs at 1

critical value, 12.84

test statistic, 23.87

▲ **Figure 3.8**

From the tables, for the 0.5% significance level and $v = 3$, the critical value of χ_3^2 is 12.84.

Since $23.87 > 12.84$, H_0 is rejected.

The data indicate that the binomial distribution is not an appropriate model for the number of marked eggs. If you look at the distribution of the data, as illustrated in Figure 3.9, you can easily see why, since it is bimodal.

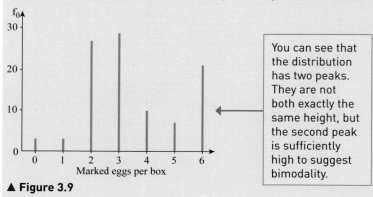

Marked eggs per box

You can see that the distribution has two peaks. They are not both exactly the same height, but the second peak is sufficiently high to suggest bimodality.

▲ **Figure 3.9**

Example 3.4	

It is generally believed that a particular genetic defect is carried by 10% of people. A new and simple test becomes available to determine whether somebody is a carrier of this defect, using a blood specimen. As part of a research project, 100 hospitals are asked to carry out this test anonymously on the next 30 blood samples they take. The results are as follows.

Number of positive tests	0	1	2	3	4	5	6	7+
Frequency, f_o	11	29	26	20	9	3	1	1

Do these figures support the model that 10% of people carry this defect, independently of any other condition, at the 5% significance level?

Solution

H_0: The model that 10% of people carry this defect is appropriate.

H_1: The model that 10% of people carry this defect is not appropriate.

The expected frequencies may be found using the binomial distribution $B(30, 0.1)$.

Number of positive tests	0	1	2	3	4	5	6	7+
Expected frequency, f_e	4.24	14.13	22.77	23.61	17.71	10.23	4.74	2.58

The calculation then proceeds as follows.

Since the expected frequency for 0 positive tests is less than 5, you need to combine this class with the class for 1 positive test. For the same reason, you also need to combine the classes for 6 and 7+ positive tests.

No. of positive tests	0, 1	2	3	4	5	6+	Total
Observed frequency, f_o	40	26	20	9	3	2	100
Expected frequency, f_e	18.37	22.77	23.61	17.71	10.23	7.32	100.01
$(f_o - f_e)$	21.63	3.23	−3.61	−8.71	−7.23	−5.32	
$(f_o - f_e)^2/f_e$	25.47	0.46	0.55	4.28	5.11	3.87	

The test statistic $X^2 = 25.47 + 0.46 + 0.55 + 4.28 + 5.11 + 3.87$

$$= 39.74$$

The degrees of freedom,

$$\nu \quad = \quad 6 \quad - \quad 0 \quad - \quad 1 \quad = \quad 5$$

number of classes	No parameters were estimated.	One degree of freedom lost because of restriction of total.

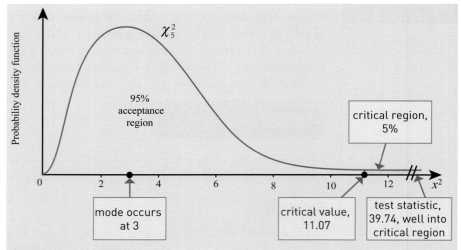

▲ **Figure 3.10**

From the tables, for 5% significance level and $v = 5$, the critical value of χ_5^2 is 11.07. Since $39.74 > 11.07$, H_0 is rejected.

The data indicate that the binomial distribution with $p = 0.1$ is not an appropriate model.

Notes

1 Although this example is like the previous one in that both used the binomial distribution as a model, the procedure is different. In this case, the given model included the information $p = 0.1$ and so you did not have to estimate the parameter p. Consequently, a degree of freedom was not lost from doing so.

2 The value of X^2 was very large in comparison with the critical value. What went wrong with the model?

3 You will find that if you use the data to estimate p, it does not work out to be 0.1 but a little under 0.07. Fewer people are carriers of the defect than was believed to be the case. If you work through the example again with the model $p = 0.07$, you will find that the fit is good enough for you to start looking at the left-hand tail of the distribution.

ACTIVITY 3.1

Carry out a goodness of fit test for the data in Example 3.4 using the p-value estimated from the data.

Goodness of fit test for the geometric distribution

| Example 3.5 | At a talent-spotting event for young football players, 100 of them take penalties until they score a goal. Data on the number of attempts required for a player to score are given below. |

No. of attempts	1	2	3	4	5	6	7	8	9	**Total**
No. of players, f_o	20	23	19	17	10	6	3	0	2	**100**

It is thought that the distribution may be modelled by the geometric distribution. The spreadsheet shows calculations for the test statistic and the p-value.

(i) Give a spreadsheet formula that could be used to calculate the test statistic from the contributions.

(ii) Use the spreadsheet output to carry out a test at the 5% significance level to determine whether the data can be modelled by the geometric distribution.

	A	B	C	D	E	F	G	H	I	J
1	Attempts	1	2	3	4	5	6	7	8	9
2	Observed frequency	20	23	19	17	10	6	3	0	2
3	Probability	0.3164557	0.21631	0.14786	0.10107	0.06908	0.04722	0.03228	0.02206	0.01508
4	Expected frequency	31.64667	21.6311	14.7858	10.1068	6.90843	4.72222	3.22785	2.20638	1.50816
5										
6	Attempts	1	2	3	4	5	6 or 7	>=8	Sum	
7	Observed frequency	20	23	19	17	10	9	2	100	
8	Probability	0.31646	0.21631	0.14786	0.10107	0.06908	0.07950	0.6972	1	
9	Expected frequency	31.65	21.63	14.79	10.11	6.91	7.95	6.97	100	
10	Contribution	4.2856	0.0866	1.2011	4.7014	1.3835	0.1387	3.5459	15.3427	
11										
12	Mean	3.16								
13	Geometric p	0.3165								
14										
15	X^2 statistic	15.3427								
16	d of f	5								
17	p-value	0.008994								

▲ Figure 3.11

(iii) Explain why the number of degrees of freedom is 5.

(iv) Explain why there is no expected frequency for 9 in the second table.

 Note

You are not told the value of p so you have to estimate it from the data. To do this, you first find the mean and then use the fact that, for a geometric distribution, $E(X) = \frac{1}{p}$.

Answers to exercises are available at www.hoddereducation.com/cambridgeextras

Solution

(i)   Sum(B10:H10).

(ii) H_0: The number of attempts can be modelled by the geometric distribution.

H_1: The number of attempts cannot be modelled by the geometric distribution.

The p-value of $0.008994 < 0.05$ so you have sufficient evidence to reject H_0 and conclude that the geometric model may not be appropriate.

(iii) There are seven classes and two restrictions, the total number of 100 and the estimated value of p, so there are $7 - 1 - 1 = 5$ degrees of freedom.

(iv) There is no expected frequency for 9 or more attempts since the expected frequency for 9 or more would be $6.97 - 2.21 = 4.76$, which is less than 5.

Note

You know from cell H9 that the expected frequency for $\geqslant 8$ is 6.97 and from cell I4 that the expected frequency for 8 is 2.21.

Testing whether data come from a given normal distribution

It is a common assumption in statistics that a set of data arises from normal distribution, for instance when conducting a t-test. How can you test, from the information in a sample, whether this assumption is justified? One method, using the χ^2 test, is shown in the example below.

Example 3.6

When an intelligence test was standardised, scores on the test were distributed normally with mean 100 and standard deviation 15. Twenty years later, it is thought that the distribution of scores may have changed.

The intelligence scores of a random sample of 40 people were measured and are given below.

92	106	91	112	106	113	125	108
103	127	110	112	120	97	115	90
119	87	114	90	88	117	119	108
103	94	104	116	97	112	103	97
86	82	114	120	115	94	110	106

Use these data to test whether the distribution is indeed still normal.

Solution

The hypotheses are:

H_0: Q has the distribution $N(100, 15^2)$

H_1: Q does not have the distribution $N(100, 15^2)$

where Q is the random variable giving an individual's test score.

One possible way of grouping these data is as follows.

Range of IQ scores	Observed frequency, f_o
−86.5	2
86.5–95.5	8
95.5–104.5	10
104.5–113.5	8
113.5–	12

You can now calculate the frequencies you would expect in these intervals under the assumption that the null hypothesis is true.

You know that Q is a random variable for which the distribution is $N(100, 15^2)$ under the null hypothesis, so that

$$P(86.5 < Q < 95.5) = P\left(\frac{86.5 - 100}{15} < Z < \frac{95.5 - 100}{15}\right)$$

$$= P(-0.9 < Z < -0.3)$$

$$= \Phi(0.9) - \Phi(0.3)$$

$$= 0.8159 - 0.6179$$

$$= 0.1980$$

and so the expected frequency for the interval 86.5–95.5 is $0.1980 \times 40 = 7.920$.

The other expected frequencies can be found by similar calculations (and the symmetry of the intervals about the hypothesised mean), resulting in the table of observed and expected frequencies below.

Range of IQ scores	Observed frequency, f_o	Expected frequency, f_e
−86.5	2	7.362
86.5–95.5	8	7.920
95.5–104.5	10	9.433
104.5–113.5	8	7.920
113.5–	12	7.362

The statistic

$$X^2 = \sum_{\text{All classes}} \frac{(f_o - f_e)^2}{f_e}$$

has an approximately χ^2 distribution under the null hypothesis. Its degrees of freedom are given by

$$\nu = \text{number of classes} - 1$$

where one degree of freedom is lost because the totals of expected and observed frequencies are equal.

Answers to exercises are available at www.hoddereducation.com/cambridgeextras

Here

$$X^2 = \frac{(2 - 7.362)^2}{7.362} + \frac{(8 - 7.920)^2}{7.920} + \frac{(10 - 9.433)^2}{9.433}$$
$$+ \frac{(8 - 7.920)^2}{7.920} + \frac{(12 - 7.362)^2}{7.362}$$

$$= 6.863$$

and $\nu = 5 - 1 = 4$.

Observe that the expected frequencies for each cell are greater than 5, as required for the χ^2 distribution to be a good approximation, so that you can use the χ^2 tables to determine the critical value; for 4 degrees of freedom at the 5% level this is 9.488. Since $6.863 < 9.488$, you can accept the null hypothesis that the sample is drawn from an underlying $N(100, 15^2)$ distribution.

 Note

It is very important to be clear exactly what the acceptance of the null hypothesis means: that it is not particularly implausible that the data seen could have arisen from random sampling of the stated normal distribution. In no sense have you confirmed that the underlying distribution does have this form, merely that it is not unreasonable to assume that it does. The same data used in a t-test to test the null hypothesis $\mu = 100$, on the assumption that the sample is drawn from a normal distribution with mean μ, leads to rejection of the null hypothesis at the 1% level.

Testing for normality without a known mean and variance

When testing for normality of the underlying distribution, in preparation for conducting a t-test, for instance, you are merely asking whether it is appropriate to assume that the underlying distribution is normal in shape; not whether it has a specific mean and variance.

Example 3.7

An experiment is conducted to determine whether people's estimates of one minute have a mean duration of one minute. Data is to be collected by asking a sample of people to say 'start' and 'stop' at times they estimate to be one minute apart. The actual time apart, in seconds, is recorded by the experimenter. A t-test is to be conducted of the hypothesis that the mean actual time apart is 60 seconds. Before this is done, a preliminary sample is taken.

The estimates (in seconds) obtained when this preliminary sample was taken are listed below.

55	40	50	53	57	61	38	29	43	52
37	57	55	56	57	48	59	40	54	53
63	58	56	48	55	58	57	56	59	55

50	60	58	51	42	47	62	57	49	43
51	42	39	56	53	53	58	51	50	55
40	38	41	55	45	61	53	53	41	53

Use these data to decide whether the assumption of normality is reasonable.

Solution

The hypotheses to be tested are:

H_0: Estimates are normally distributed.

H_1: Estimates are not normally distributed.

With these data, start by calculating the sample mean and the usual sample estimate of the population standard deviation.

$$\bar{x} = 51.1 \quad \text{and} \quad s = \sqrt{\frac{\sum (x - \bar{x})^2}{n - 1}} = 7.564$$

Now use these estimated parameters to calculate the expected frequencies: that is, test the fit of the data to the normal distribution $N(51.1, 7.564^2)$.

The data must now be grouped. One possible grouping is shown below.

Estimates (seconds)	Observed frequency, f_o
−41.5	10
41.5–45.5	5
45.5–49.5	4
49.5–53.5	14
53.5–57.5	16
57.5–	11

The expected frequencies for these groups can be calculated as set out below.

Class	Upper class boundary	Standardised value z	$P(Z < z)$	Probability for class	Expected frequency, f_e
−41.5	41.5	−1.2692	0.1022	0.1022	6.132
41.5–45.5	45.5	−0.7404	0.2295	0.1273	7.638
45.5–49.5	49.5	−0.2115	0.4162	0.1867	11.202
49.5–53.5	53.5	0.31729	0.6244	0.2082	12.492
53.5–57.5	57.5	0.84611	0.8012	0.1768	10.608
57.5–	∞	∞	1	0.1988	11.928

→

The statistic is

$$X^2 = \sum_{\text{All groups}} \frac{(f_o - f_e)^2}{f_e}$$

$$= \frac{(10 - 6.132)^2}{6.132} + \frac{(5 - 7.638)^2}{7.638} + \frac{(4 - 11.202)^2}{11.202} + \frac{(14 - 12.492)^2}{12.492}$$

$$+ \frac{(16 - 10.608)^2}{10.608} + \frac{(11 - 11.928)^2}{11.928}$$

$$= 2.440 + 0.911 + 4.630 + 0.182 + 2.741 + 0.072 = 10.976.$$

To calculate the degrees of freedom, recall that you have used the data to estimate two parameters (the mean and the standard deviation) for the distribution. This means that both the observed and expected frequencies must give the same total frequency, the same sample mean and the same sample estimate of the standard deviation. These three restrictions on the possible frequencies in each class reduce the degrees of freedom by three from the number of classes.

$$\nu \quad = \quad 6 \quad - \quad 1 \quad - \quad 2 \quad = \quad 3$$

| number of classes | total frequency fixed | number of parameters estimated from the data |

With 3 degrees of freedom, the critical value at the 5% level for the χ^2 distribution is 7.815.

Since $10.976 > 7.815$, you reject the null hypothesis that the data were drawn from a normally distributed population, and conclude that it would not be appropriate to use a t-test for assessing the hypothesis $\mu = 60$.

> Explain why, despite this result, you could use a normal test.

As before, from the values of $\frac{(f_o - f_e)^2}{f_e}$ for each cell calculated above you can see which differences between observed and expected frequencies are important, in the sense of making a large contribution to the χ^2 statistic. The observed frequency of 10 in the first cell is itself substantially higher than the expected frequency of 6.132, but this excess of low estimates also brings down the estimate of the mean to the lower end of the main peak of the distribution. Hence a class above that containing the estimated mean has a significantly higher observed than expected frequency and the class below that containing the estimated mean a significantly lower observed than expected frequency.

Note

There is a certain amount of arbitrariness in the grouping of data that precedes the 'goodness-of-fit' test; you want to ensure that there are enough classes to discriminate between different distributions, but that each is wide enough to have an expected frequency of at least 5. There is no need to choose constant class widths, and, in fact, it would be wise to have narrower classes where the expected distribution has the greatest density. Picking class widths so that the expected frequencies are all about 8–12 is a reasonable rule of thumb.

Testing goodness of fit with other continuous distributions

You can also use the χ^2 test for the goodness of fit of a set of data to other continuous underlying distributions, not just the normal, as the example below illustrates.

Example 3.8

It is suggested that the time intervals between arrivals of passengers at a bus stop can be modelled by an exponential distribution, provided that buses appear sufficiently frequently and unpredictably for passengers simply to turn up independently at random.

The time intervals between 36 successive arrivals at a bus stop were measured (in seconds) and are recorded below.

60	43	25	25	31	37	23	23	14
13	36	3	63	50	33	18	60	52
6	38	41	33	52	28	36	30	42
16	10	55	161	21	1	3	14	13

Use these data to investigate the suggestion.

Solution

The hypotheses to be tested are as follows.

H_0: The time intervals between passenger arrivals are exponentially distributed.

H_1: The time intervals between passenger arrivals are not exponentially distributed.

A grouped frequency distribution for the sample looks like this.

Time interval (nearest second)	Observed frequency, f_o
0–9	4
10–19	7
20–29	6
30–39	8
40–49	3
50–59	4
60–69	3
70–79	1

→

Answers to exercises are available at www.hoddereducation.com/cambridgeextras

To calculate the expected frequencies, you need to use the probability density function for the exponential distribution. This is given by

$$f(t) = \lambda e^{-\lambda t}$$

The mean of this data is 33.58.

The mean of the distribution is $\frac{1}{\lambda}$ so you can estimate λ from the sample mean as $\frac{1}{\bar{t}} = 0.029\,78$.

The probability that an exponentially distributed variable lies between a and b is

$$\int_a^b \lambda e^{-\lambda t} dt = [-e^{-\lambda t}]_a^b = e^{-\lambda a} - e^{-\lambda b}$$

so the expected frequencies can be calculated, for example for the class interval 20–29 (that is, with class boundaries 19.5 and 29.5) as

$$36 \times (e^{-0.02978 \times 19.5} - e^{-0.02978 \times 29.5}) = 36 \times 0.144\,10 = 5.1874.$$

The complete set of expected frequencies is as follows:

Time interval (nearest second)	Probability	Expected frequency, f_e
0–9	0.246 39	8.8700
10–19	0.194 07	6.9867
20–29	0.144 10	5.1874
30–39	0.106 99	3.8515
40–49	0.079 44	2.8597
50–59	0.058 98	2.1232
60–69	0.043 79	1.5764
70–	0.126 25	4.5451

Note that the expected frequency is calculated for the entire interval 70–∞, not just the interval 70–79 in which the maximum of the actual data lies. In conducting a χ^2 test, it is essential that the intervals for which the expected frequencies are calculated cover the whole range of the theoretical distribution, not just the range of the actual data. An alternative way of doing this would have been to add an extra interval 80–∞, with observed frequency 0. However, in this example, the expected frequency would have been considerably less than 5, so the figures used to calculate the final X^2 statistic would have been the same. These expected frequencies are not all greater than 5, so the class boundaries need to be redrawn. One way of doing this is as follows.

Time interval (nearest second)	Observed frequency, f_o	Expected frequency, f_e
0–9	4	8.8700
10–19	7	6.9867
20–29	6	5.1874
30–49	11	6.7112
50–	8	8.2447

The statistic is

$$X^2 = \sum_{\text{All groups}} \frac{(f_o - f_e)^2}{f_e}$$

$$= \frac{(4 - 8.8700)^2}{8.8700} + \frac{(7 - 6.9867)^2}{6.9867} + \frac{(6 - 5.1874)^2}{5.1874}$$

$$+ \frac{(11 - 6.7112)^2}{6.7112} + \frac{(8 - 8.2447)^2}{8.2447}$$

$$= 5.549$$

parameter λ estimated from data

and the degrees of freedom are

classes

sum of frequencies fixed

$$\nu \quad = \quad 5 \quad - \quad 1 \quad - \quad 1 \quad = \quad 3$$

The critical value at the 5% level for 3 degrees of freedom is 7.815.

Since $5.549 < 7.815$ you accept the null hypothesis that the underlying distribution from which the data are drawn is exponential.

The left-hand tail

The χ^2 test is conducted as a one-tailed test, looking to see if the test statistic gives a value to the right of the critical value, as in the previous examples.

However, examination of the left-hand tail also gives information. In any modelling situation you would expect there to be some variability. Even when using the binomial to model a clear binomial situation, like the number of heads obtained in throwing a coin a large number of times, you would be very surprised if the observed and expected frequencies were identical. The left-hand tail may lead you to wonder whether the fit is too good to be credible. The following is a very famous example of the Poisson distribution.

Death from horse kicks

For a period of 20 years in the 19th century, data were collected of the annual number of deaths caused by horse kicks per army corps in the Prussian army.

No. of deaths	0	1	2	3	4
No. of corps–years, f_o	109	65	22	3	1

These data give a mean of 0.61. The variance of 0.6079 is almost the same, suggesting that the Poisson model may be appropriate.

The Poisson distribution (0.61) gives these figures:

No. of deaths	0	1	2	3	4 or more
No. of corps–years, f_o	108.7	66.3	20.2	4.1	0.7

This looks so close that there seems little point in using a test to see if the data will fit the distribution. However, proceeding to the test, and remembering to combine classes to give expected values of at least 5, the null and alternative hypotheses are as follows.

No. of deaths	0	1	2 or more
Observed frequency, f_o	109	65	26
Expected frequency, f_e	108.7	66.3	25
$(f_o - f_e)$	0.3	−1.3	1
$(f_o - f_e)^2/f_e$	0.001	0.025	0.040

H_0: Deaths from horse kicks can be modelled by a Poisson distribution.

H_1: Deaths from horse kicks cannot be modelled by a Poisson distribution.

The test statistic $X^2 = 0.001 + 0.025 + 0.040$

$$= 0.066$$

The degrees of freedom are given by

the total frequency imposes one restriction

number of classes

one parameter was estimated

$$\nu \quad = \quad 3 \quad - \quad 1 \quad - \quad 1 \quad = \quad 1$$

The critical value of χ^2 for 1 degree of freedom at the 5% significance level is 3.841.

As $0.066 < 3.841$ it is clear that the null hypothesis, that the Poisson distribution is an appropriate model, should be accepted.

The tables relating to the left-hand tail of the χ^2 distribution give critical values for this situation. For example, the value for 95% significance level for $\nu = 1$ is 0.0039. This means that if the null hypothesis is true, you would expect a value for X^2 less than 0.0039 from no more than 5% of samples.

> Looking at the expected values of the Poisson distribution in comparison with the observed values suggests that the fit is very good indeed. Is it perhaps suspiciously good? Might the data have been fixed?

In this case, the test statistic is $X^2 = 0.066$. This is greater than the 95% critical value, and so you can conclude that a fit as good as this will occur with more than 1 sample in 20. That may well help allay your suspicions.

If your test statistic does lie within the left-hand critical region, you should check the data to ensure that the figures are genuine and that all the procedures have been carried out properly. There are three situations you should particularly watch out for:

» The model was constructed to fit a set of data. It is then being tested by seeing how well it fits the same data. Once the model is determined, new data should be used to test it.

» Some of the data have been omitted in order to produce a better fit.

» The data are not genuine.

Although looking at the left-hand tail of the χ^2 distribution may make you suspicious of the quality of the data, it does not provide a formal hypothesis test that the data are not genuine. Thus the term **critical region** is not really appropriate to this tail; **warning region** would be better.

The χ^2 test is a **distribution-free test**: this means that there are no modelling assumptions associated with the test itself. This has the advantage that the test can be widely used, but, on the other hand, it is not a very sensitive test. The non-parametric tests that you will meet in Chapter 4 are further examples of distribution–free tests; one can carry out these tests without knowing anything about the underlying distributions.

> If you obtain a p-value for a goodness of fit test, you need to take a little care interpreting it. The lower the p-value, the less likely it is that the sample could have been drawn from a distribution for which the null hypothesis is true; conversely, the higher the p-value, the more likely it is that that the sample could have been drawn from such a distribution. Quite often you will want to show that a certain distribution is appropriate and so you will associate a high p-value with a successful outcome. By contrast, in many other hypothesis tests, you are hoping to show that the null hypothesis is false, and so something special has happened; in such cases, a low p-value can be regarded as a successful outcome.

Exercise 3B

1 Find the expected frequencies of $0, 1, 2, 3, \geq 4$ successes if 80 observations are taken of a binomial random variable with $n = 20$ and $p = 0.09$.

2 You are given that $X \sim$ Poisson (1.6). Find the expected frequencies of $x = 0, 1, 2, 3, 4, \geq 5$ if 60 observations of X are taken.

3 A typist makes mistakes from time to time in a 200-page book. The numbers of pages with mistakes are as follows.

Mistakes	0	1	2	3	4	5
Pages	18	62	84	30	5	1

(i) Test at the 5% significance level whether the Poisson distribution is an appropriate model for these data.

(ii) What factors would make it other than a Poisson distribution?

Answers to exercises are available at www.hoddereducation.com/cambridgeextras

M 4 A biologist crosses two pure varieties of plant, one with pink flowers, the other white. The pink is dominant so that the flowers of the second generation should be in the ratio pink : white = 3 : 1.

He plants the seeds in batches of 5 in 32 trays and counts the numbers of plants with pink and with white flowers in each tray.

White flowers	0	1	2	3	4	5
Frequency	9	12	7	2	1	1

(i) What distribution would you expect for the number of plants with white flowers?

(ii) Use these figures to test at the 2.5% significance level whether the distribution is that which you expected.

M 5 An examination board is testing a multiple-choice question. They get 100 students to try the question and their answers are as follows.

Choice	A	B	C	D	E
Frequency	32	18	10	28	12

Are there grounds, at the 10% significance level, for the view that the question was so hard that the students guessed the answers at random?

M 6 A student on a geography field trip has collected data on the size of rocks found on a scree slope. The student counts the number of large rocks (that is, heavier than a stated weight) found in a 2 m square at the top, middle and bottom of the slope.

	Top	Middle	Bottom
Number of large rocks	5	10	18

(i) Test at the 5% significance level whether these data are consistent with the hypothesis that the size of rocks is distributed evenly on the scree slope.

(ii) What does your test tell the student about the theory that large rocks will migrate to the bottom of the slope?

M 7 A university student working in a small coastal hotel in the summer holidays looks at the records for the previous holiday season of 30 weeks. She records the number of days in each week on which the hotel had to turn away visitors because it was full. The data she collects are as follows.

Number of days visitors turned away	0	1	2	3	4	5+	Total
Number of weeks	11	13	4	1	1	0	30

(i) Calculate the mean and variance of the data.

(ii) The student thinks that these data can be modelled by the binomial distribution. Carry out a test at the 5% significance level to see if the binomial distribution is a suitable model.

(iii) What other distribution might be used to model these data? Give your reasons.

8 Li is writing a book. Every so often she uses the spell-check facility in her word-processing software, and, for interest, records the number of mistakes she has made on each page. In the first 20 pages, the results were as follows.

No. of mistakes/page	0	1	2	3	4+	**Total**
Frequency	9	6	4	1	0	**20**

(i) Explain why it is not possible to use the χ^2 test on these data to decide whether the occurrence of spelling mistakes may be modelled by the Poisson distribution.

In the next 30 pages Li's figures are as follows.

No. of mistakes/page	0	1	2	3	4+	**Total**
Frequency	14	7	7	0	2	**30**

(ii) Use the combined figures, covering the first 50 pages, to test whether the occurrence of Li's spelling mistakes may be modelled by the Poisson distribution. Use the 5% significance level.

(iii) If the distribution really is Poisson, what does this tell you about the incidence of spelling mistakes? Do you think this is realistic?

9 In a survey of five towns, the population of the town and the number of petrol-filling stations were recorded as follows.

Town	Population (to nearest 10 000)	Number of filling stations
A	4	22
B	3	16
C	7	35
D	6	27
E	12	60
Totals	**32**	**160**

An assistant researcher, who wanted to find out whether the petrol stations were evenly distributed between the towns, performed a χ^2 test on the number of filling stations, with a null hypothesis that there was no difference in the number of filling stations in each town. She found that her X^2 value was 36.69. Without repeating her calculation, state with reasons what her conclusion was.

Answers to exercises are available at www.hoddereducation.com/cambridgeextras

M 10 Fifty-five people are asked to estimate a one-metre length, by marking off their estimate on a blank straight edge. The actual length marked off is then recorded.

The results are given below, in centimetres.

112	109	89	110	116	99	109	120	132
80	95	101	107	142	110	111	76	89
100	103	132	117	121	112	110	126	105
98	108	80	87	97	116	126	104	110
128	103	88	118	72	77	87	117	126
114	115	118	120	98	117	81	91	107
115								

Use these data to carry out a χ^2 test for goodness of fit to the distribution $N(100, \sigma^2)$, where σ^2 is to be estimated from the data.

11 The number of goals scored by a certain football team was recorded for each of 100 matches, and the results are summarised in the following table.

Number of goals	0	1	2	3	4	5	6 or more
Frequency	12	16	31	25	13	3	0

Fit a Poisson distribution to the data, and test its goodness of fit at the 5% significance level.

Cambridge International AS & A Level Further Mathematics
9231 Paper 21 Q8 November 2015

12 Drinking glasses are sold in packs of 4. The manufacturer conducts a survey to assess the quality of the glasses. The results from a sample of 50 randomly chosen packs are summarised in the following table.

Number of perfect glasses	0	1	2	3	4
Number of packs	1	3	10	17	19

Fit a binomial distribution to the data and carry out a goodness of fit test at the 10% significance level.

Cambridge International AS & A Level Further Mathematics
9231 Paper 23 Q8 November 2012

13 The numbers of a particular type of laptop computer sold by a store on each of 100 consecutive Saturdays are summarised in the following table.

Number sold	0	1	2	3	4	5	6	7	$\geqslant 8$
Number of Saturdays	7	20	39	16	14	2	1	1	0

Fit a Poisson distribution to the data and carry out a goodness of fit test at the 2.5% significance level.

Cambridge International AS & A Level Further Mathematics
9231 Paper 22 Q8 November 2014

14 Applicants for a national teacher training course are required to pass a mathematics test. Each year, the applicants are tested in groups of 6 and the number of successful applicants in each group is recorded. The overall proportion of successful applicants has remained constant over the years and is equal to 60% of the applicants. The results from 150 randomly chosen groups are shown in the following table.

Number of successful applicants	0	1	2	3	4	5	6
Numbers of groups	1	3	25	51	38	30	2

Test, at the 5% significance level, the goodness of fit of the distribution $B(6, 0.6)$ for the number of successful applicants in a group.

Cambridge International AS & A Level Further Mathematics
9231 Paper 21 Q9 June 2016

15 The number of visitors arriving at an art exhibition is recorded for each 10-minute period of time during the ten hours that it is open on a particular day. The results are as follows.

Number of visitors in a 10-minute period	0	1	2	3	4	5	6	7	8	$\geqslant 9$
Number of 10-minute periods	2	2	12	8	11	13	4	7	1	0

(i) Calculate the mean and variance for this sample and explain whether your answers support a suggestion that a Poisson distribution might be a suitable model for the number of visitors in a 10-minute period.

(ii) Use an appropriate Poisson distribution to find the two expected frequencies missing from the following table.

Number of visitors in a 10-minute period	0	1	2	3	4	5	6	7	8	$\geqslant 9$
Expected number of 10-minute periods	1.10		8.79		11.72	9.38	6.25	3.57	1.79	1.28

(iii) Test, at the 10% significance level, the goodness of fit of this Poisson distribution to the data.

Cambridge International AS & A Level Further Mathematics
9231 Paper 21 Q9 November 2016

Answers to exercises are available at www.hoddereducation.com/cambridgeextras

16 Each of 200 identically biased dice is thrown repeatedly until an even number is obtained. The number of throws, x, needed is recorded and the results are summarised in the following table.

x	1	2	3	4	5	6	$\geqslant 7$
Frequency	126	43	22	3	5	1	0

State a type of distribution that could be used to fit the data given in the table above.

Fit a distribution of this type in which the probability of throwing an even number for each die is 0.6 and carry out a goodness of fit test at the 5% significance level.

For each of these dice, it is known that the probability of obtaining a 6 when it is thrown is 0.25. Ten of these dice are each thrown 5 times. Find the probability that at least one 6 is obtained on exactly 4 of the 10 dice.

Cambridge International AS & A Level Further Mathematics
9231 Paper 21 Q11 June 2015

17 A sample of 216 observations of the continuous random variable X was obtained and the results are summarised in the following table.

Interval	$0 \leqslant x < 1$	$1 \leqslant x < 2$	$2 \leqslant x < 3$	$3 \leqslant x < 4$	$4 \leqslant x < 5$	$5 \leqslant x < 6$
Observed frequency	1	3	15	31	59	107

It is suggested that these results are consistent with a distribution having probability density function f given by

$$f(x) = \begin{cases} kx^2 & 0 \leqslant x < 6, \\ 0 & \text{otherwise,} \end{cases}$$

where k is a positive constant. The relevant expected frequencies are given in the following table.

Interval	$0 \leqslant x < 1$	$1 \leqslant x < 2$	$2 \leqslant x < 3$	$3 \leqslant x < 4$	$4 \leqslant x < 5$	$5 \leqslant x < 6$
Expected frequency	1	7	a	b	c	91

(i) Show that $a = 19$ and find the values of b and c.

(ii) Carry out a goodness of fit test at the 10% significance level.

Cambridge International AS & A Level Further Mathematics
9231 Paper 21 Q8 November 2011

18 A random sample of 200 observations of the continuous random variable
X was taken and the values are summarised in the following table.

Interval	$1 \leqslant x < 2$	$2 \leqslant x < 3$	$3 \leqslant x < 4$	$4 \leqslant x < 5$
Observed frequency	63	45	32	25
Interval	$5 \leqslant x < 6$	$6 \leqslant x < 7$	$7 \leqslant x < 8$	
Observed frequency	22	7	6	

It is required to test the goodness of fit of the distribution with
probability density function f given by

$$f(x) = \begin{cases} \dfrac{1}{x \ln 8} & 1 \leqslant x < 8, \\ 0 & \text{otherwise.} \end{cases}$$

The relevant expected frequencies, correct to 2 decimal places, are given
in the following table.

Interval	$1 \leqslant x < 2$	$2 \leqslant x < 3$	$3 \leqslant x < 4$	$4 \leqslant x < 5$
Observed frequency	66.67	p	27.67	q
Interval	$5 \leqslant x < 6$	$6 \leqslant x < 7$	$7 \leqslant x < 8$	
Observed frequency	17.54	14.83	12.84	

Show that $p = 39.00$, correct to 2 decimal places, and find the value of q.

Carry out a goodness of fit test at the 5% significance level.

Cambridge International AS & A Level Further Mathematics
9231 Paper 21 Q9 June 2014

KEY POINTS

1   The chi-squared test for a contingency table is used to test whether
    the variables in an $m \times n$ contingency table are independent. The steps
    are as follows.

    (i)   The null hypothesis is that the variables are independent, the
          alternative is that they are not.

    (ii) Calculate the marginal (row and column) totals for the table.

    (iii) Calculate the expected frequency in each cell.

    (iv) The X^2 statistic is $\sum \dfrac{(f_o - f_e)^2}{f_e}$ where f_o is the observed
          frequency and f_e is the expected frequency in each cell.

    (v)   The degrees of freedom, ν, for the test is $(m - 1)(n - 1)$ for an
          $m \times n$ table.

Answers to exercises are available at www.hoddereducation.com/cambridgeextras

(vi) Read the critical value from the χ^2 tables for the appropriate degrees of freedom and significance level. If X^2 is less than the significance level, the null hypothesis is accepted; otherwise it is rejected.

(vii) If two variables are not independent, you say that there is an association between them.

2 A goodness of fit test is used to test whether a distribution models a situation. The steps are as follows.

(i) Select your model, binomial, Poisson, etc.

(ii) Set up null and alternative hypotheses and choose the significance level.

(iii) Collect data. Record the observed frequency for each outcome.

(iv) Calculate the expected frequencies arising from the model.

(v) Check that the expected frequencies are all at least 5. If not, combine classes.

(vi) Calculate the test statistic.

The X^2 statistic is $\sum \dfrac{(f_o - f_e)^2}{f_e}$.

(vii) Find the degrees of freedom, ν, for the test using the formula

ν = number of classes − number of estimated parameters − 1

where the number of classes is counted after any necessary combining has been done.

(viii) Read the critical value from the χ^2 tables for the appropriate degrees of freedom and significance level. If X^2 is less than the significance level, the null hypothesis is accepted; otherwise it is rejected.

(ix) Draw conclusions from the test — state what the test tells you about the model.

LEARNING OUTCOMES

Now you have finished this chapter, you should be able to

- use a χ^2-test, with the appropriate degrees of freedom, for independence in a contingency table

- fit a theoretical distribution, as prescribed by a given hypothesis, to given data

- carry out a goodness of fit test for a discrete distribution such as a binomial distribution or a Poisson distribution

- carry out a goodness of fit test for a normal distribution

- carry out a goodness of fit test for a continuous random variable given its probability density function.

4 Non-parametric tests

> Whenever a large sample of chaotic elements are taken in hand and marshalled in the order of their magnitude, an unsuspected and most beautiful form of regularity proves to have been latent all along.
> *Sir Francis Galton (1822–1911)*

Gandhi's popularity was measured by the size of the crowds he drew. These days it is more common for opinion pollsters to ask people questions like the one on the next page.

4.1 Single-sample non-parametric tests

> ! In order to use the *t*-distribution, the parent population must be normally distributed. If you also know the population variance, then the normal distribution itself can be used instead of the *t*-distribution. If you do not know the distribution of the parent population, then you cannot use either of the distributions (unless, of course, the sample size is large, in which case you can make use of the central limit theorem and so use the normal distribution).

4

agree strongly	agree	inclined to agree	have no opinion	inclined to disagree	disagree	disagree strongly

'The President is doing the best possible job, in the circumstances.' Please choose one of the following responses.

You will probably recognise this as the sort of question that is asked by opinion pollsters. However, surveys of people's attitudes are not just undertaken on political issues: market researchers for businesses, local authorities, psychologists and pressure groups, for instance, are all interested in what we think about a very wide variety of issues.

You are going to test the hypothesis:

The President's performance is generally disapproved of.

The question above was asked of a group of twelve 17–year-olds and each reply recorded as a number from 1 to 7, where 1 indicates 'agree strongly', and 7 indicates 'disagree strongly'. This method of recording responses gives a **rating scale** of attitudes to the President's performance. The data obtained are shown below.

3   6   7   4   3   4   7   3   5   6   5   6

What do these data indicate about the validity of the hypothesis in the population from which the sample was drawn?

You should recognise this question as similar to those you asked when conducting t-tests. You want to know whether attitudes in the population as a whole are centred around the neutral response of '4' or show lower approval in general. That is, you want a test of location of the sample: one which decides what values are taken, on average, in the population.

You could not use a t-test here to decide whether the mean of the underlying distribution equals 4 because the response variable is clearly not normally distributed: it only takes discrete values from 1 to 7 (and the sample size is small).

This chapter looks at some tests of location that are valid even for small samples without the strict distributional assumptions required by the t-test, and which are therefore more widely applicable.

> **Note**
>
> Tests such as the Wilcoxon test are known as 'non-parametric' since they do not rely on the parameters of a probability distribution.

The sign test

One very simple way of handling the sort of data you have here is to make the following hypotheses.

H_0: People are equally likely to agree or disagree with the statement.

H_1: People are more likely to disagree than agree.

An opinion has been expressed by ten people (the two whose response is coded as '4' expressed no opinion); in carrying out the test, it is assumed that they constitute a random sample from the population. If the null hypothesis is true, then each of these people will, independently, agree or disagree with the statement with probability $\frac{1}{2}$. The number agreeing, X, will therefore have a binomial distribution $B(10, \frac{1}{2})$.

To test the (one-tailed) hypothesis at the 5% level you are looking for the greatest value of x for which $P(X \leqslant x) \leqslant 0.05$.

Using a calculator for $B(10, \frac{1}{2})$,

$$P(X = 0) = 0.0010$$

$$P(X = 1) = 0.0098$$

$$P(X = 2) = 0.0439$$

so $P(X \leqslant 1) = 0.0107$

and $P(X \leqslant 2) = 0.0547.$

> To find $P(X \leqslant 1)$ you add $P(X = 0)$ and $P(X = 1)$. You get 0.0108, but working to more decimal places and then rounding gives 0.0107, which is correct to 4 d.p.

This shows that this critical value is $x = 1$ and so the critical region for the test is $[0, 1]$.

In this example, three of those responding agree with the statement, so you accept the null hypothesis that people are equally likely to agree or disagree with the statement. Thus there is insufficient evidence to suggest that the President's performance is generally disapproved of.

This is an example of the **sign test**. It has the advantage of great simplicity, is useful in many circumstances and can be decisive. Because the calculation required is so quick to carry out, the sign test is often useful in an initial exploration of a set of data, and may indeed be all that is necessary.

However, there is information in the sample that is ignored in the sign test: nobody in the sample *strongly* agreed with the statement while, of the seven who disagreed, two did so strongly. Later in this chapter, you will meet the **single-sample Wilcoxon signed-rank test**; while slightly more complicated than the sign test, it takes such extra information into account.

Example 4.1

An ornithologist records the numbers of house martins (a bird species) per day flying over a headland. Her observations cover a sample of 15 days during the migration season of this year.

  281 214 268 290 271 207 402 480 400 423 350 366 355 280 368

Records show that for several years the median number per day has been 270.5. Use a sign test to examine, at the 5% significance level, whether there has been an increase in the numbers of house martins flying over the headland during the migration season of this year.

Solution

H_0: The median number of house martins is equal to 270.5.

H_1: The median number of house martins is greater than 270.5.

Answers to exercises are available at www.hoddereducation.com/cambridgeextras

If the null hypothesis is true, then the probability that the number of house martins flying over the headland each day is greater than 270.5 is $\frac{1}{2}$; similarly, the probability that it is less than 270.5 is also $\frac{1}{2}$. The number greater than 270.5, X, will therefore have a binomial distribution B(15, $\frac{1}{2}$).

$$P(X = 15) = 0.0000$$

$$P(X = 14) = 0.0005$$

$$P(X = 13) = 0.0032$$

$$P(X = 12) = 0.0139$$

$$P(X = 11) = 0.0417$$

So $P(X \geqslant 12) = 0.0000 + 0.0005 + 0.0032 + 0.0139$

$$= 0.0176 \leqslant 0.05$$

but $P(X \geqslant 11) = 0.0592 > 0.05$

> This is a one-tailed test involving the upper tail, so you need to find the upper tail critical value.

So the critical region is [12, 13, 14, 15].

On 12 of the 15 days, there were more than 270.5 house martins.

12 lies in the critical region, so there is sufficient evidence to suggest that there has been an increase in the numbers of house martins flying over the headland this year.

| Example 4.2 |

It is known from previous trials that with a fully charged battery, a vacuum cleaner will be able to clean for a median time of 27.35 minutes before the battery runs out. The manufacturer is testing a new type of battery to see whether the cleaning time is different. The cleaning times for a random sample of 9 occasions with the new battery (rounded to 1 decimal place) are as follows.

25.4 28.3 26.7 29.2 30.6 28.7 27.2 27.6 29.6

Carry out a sign test to examine, at the 5% significance level, whether there has been a change in the average cleaning time.

Solution

H_0: The median cleaning time is equal to 27.35 minutes.

H_1: The median cleaning time is different from 27.35 minutes.

If the null hypothesis is true, then the probability that the cleaning time is greater than 27.35 minutes is $\frac{1}{2}$; similarly, the probability that it is less than 27.35 minutes is also $\frac{1}{2}$. The number of cleaning times less than 27.35, X, will therefore have a binomial distribution B(9, $\frac{1}{2}$).

$$P(X = 0) = 0.0020$$

$$P(X = 1) = 0.0176$$

$$P(X = 2) = 0.0703$$

$$P(X = 3) = 0.1641$$

> This is a two-tailed test, so you need to work out both the lower and upper critical values. However, once you have worked out the lower value, you can simply subtract this (2) from $n = 9$ to get the upper value.

So $P(X \leqslant 1) = 0.0020 + 0.0176 = 0.0195 \leqslant 0.025$

but $P(X \leqslant 2) = 0.0898 > 0.025$

> To find $P(X \leqslant 1)$ you add $P(X = 0)$ and $P(X = 1)$. You get 0.0196, but working to more decimal places and then rounding gives 0.0195, which is correct to 4 d.p.; similarly for $P(X \leqslant 2)$.

So the critical region is [0, 1, 8, 9].

On 3 of the 9 trials, the time was less than 27.35.

3 does not lie in the critical region so there is insufficient evidence to suggest that the median cleaning time is different from 27.35.

Example 4.3

The median time it takes to cure a particular illness using the standard drug treatment is 23.5 days. A new drug is being tested to see if it reduces the treatment time. A random sample of 50 patients with the illness is selected and the patients are given the new drug. Of these 50 people, 33 are cured in less than 23.5 days.

Carry out a sign test to examine, at the 10% significance level, whether there is a reduction in the median time required to cure the illness when using the new drug.

Solution

H_0: The median time to cure the illness is equal to 23.5 days.

H_1: The median time to cure the illness is less than 23.5 days.

If the null hypothesis is true, then the probability that the time to cure the illness is greater than 23.5 days is $\frac{1}{2}$; similarly, the probability that the time is less than 23.5 days is also $\frac{1}{2}$. The number of times that are less than 23.5, X, will therefore have a binomial distribution $B(50, \frac{1}{2})$.

Because the sample size is large, you can use a normal approximation to carry out the test.

The mean $= np = 50 \times \frac{1}{2} = 25$ and the variance $= npq = 50 \times \frac{1}{2} \times \frac{1}{2} = 12.5$

$$P(X \geqslant 33) = 1 - \Phi\left(\frac{32.5 - 25}{\sqrt{12.5}}\right)$$

> Because we are approximating the binomial distribution (which is discrete) with the normal, we need to use a continuity correction, so we use 32.5 rather than 33.

$$= 1 - \Phi(2.121)$$

$$= 1 - 0.9830$$

$$= 0.0170$$

$0.0170 < 10\%$ so reject H_0. There is sufficient evidence to suggest that the median time to cure the illness is less than 23.5.

Answers to exercises are available at www.hoddereducation.com/cambridgeextras

The Wilcoxon signed-rank test on a sample median

You may recall that in the opening example in this chapter, it was mentioned that there was information in the sample that is ignored in the sign test: nobody in the sample *strongly* agreed with the statement about the President while, of the seven who disagreed, two did so strongly. There is a different test that does take this type of information into account. This test is the **Wilcoxon signed-rank test** on a sample median. The only assumption that this test requires is that *the data is distributed symmetrically*.

You will see how to carry out this test on the data in Example 4.1. Here is a reminder of this example.

> An ornithologist records the numbers of house martins (bird species) per day flying over a headland. Her observations cover a sample of 15 days during the migration season of this year.
>
> 281   214   268   290   271   207   402   480   400   423
> 350   366   355   280   368
>
> Records show that for several years the median number per day has been 270.5.

You will now use a Wilcoxon signed-rank test to examine, at the 5% significance level, whether there has been an increase in the numbers of house martins flying over the headland during the migration season of this year.

The hypotheses that you use for this test are exactly the same as for the sign test, namely:

H_0: The median number of house martins is equal to 270.5.

H_1: The median number of house martins is greater than 270.5.

> This is a one-tailed test involving the upper tail.

The Wilcoxon test takes into account not only whether the data values are less than or greater than the hypothesised median, but also how far above or below it they are. However, it does not use the actual data values themselves, but instead the ranks of their distances from the hypothesised median of the population, in this case 270.5.

For these data, this gives the following results.

> The ratings are ranked according to their absolute differences from the median.

Number, h	$h - 270.5$	$\lvert h - 270.5 \rvert$	Rank
281	10.5	10.5	4
214	−56.5	56.5	6
268	−2.5	2.5	2
290	19.5	19.5	5
271	0.5	0.5	1
207	−63.5	63.5	7
402	131.5	131.5	13

Number, h	$h - 270.5$	$|h - 270.5|$	Rank
480	209.5	209.5	15
400	129.5	129.5	12
423	152.5	152.5	14
350	79.5	79.5	8
366	95.5	95.5	10
355	84.5	84.5	9
280	9.5	9.5	3
368	97.5	97.5	11

Suppose that the assumption of symmetry is correct, and the null hypothesis that the median is 270.5 is true. Then you would expect a rating of, for example, 271 to come up as often as a rating of 270, a rating of 272 to come up as often as a rating of 269, a rating of 273 to come up as often as a rating of 268, and so on. In other words, ratings at each distance from the supposed median of 270.5 should be equally likely to be above or below that median.

To test whether the data support this, the next step is to calculate the sum of the ranks of the ratings above 270.5 and below 270.5 and compare these with the total sum of the ranks.

$$\text{Sum of ranks of ratings above } 270.5 = 4 + 5 + 1 + 13 + 15 + 12 + 14$$
$$+ 8 + 10 + 9 + 3 + 11 = 105$$

$$\text{Sum of ranks of ratings below } 270.5 = 6 + 2 + 7 = 15$$

$$\text{Total sums of ranks} = 105 + 15 = 120$$

Note that because of the way in which the ratings were ranked, the total sum of ranks must be equal to the sum of the numbers from 1 to 15:

> You will probably have met the formula, which states that the sum of the first n natural numbers is $\frac{1}{2}n(n+1)$.

$$1 + 2 + \ldots + 15 = \frac{1}{2} \times 15 \times (15 + 1) = 120$$

If the null hypothesis is true, you may expect the sum of the ranks of ratings above 270.5 to be approximately equal to the sum of the ranks of ratings below 270.5. This means that each would be about half of 120; that is, 60. The fact that, for these data, the sum of ranks of ratings above 270.5 is much more than this, and the sum of ranks of ratings below 270.5 is correspondingly less, implies that either there were more days when over 270.5 birds flew over or that on days when over 270.5 birds flew over, there were a lot more flying over than previously. Actually, both of these are true of the data.

In order to conduct a hypothesis test, you need to know the critical values of the test statistic. Tables of the critical values for the Wilcoxon test are available, and a section of one is shown in Figure 4.1 on the next page.

Critical values of T

	Level of significance			
One-tailed	0.05	0.025	0.01	0.005
Two-tailed	0.1	0.05	0.02	0.01
$n = 6$	2	0		
7	3	2	0	
8	5	3	1	0
9	8	5	3	1
10	10	8	5	3
11	13	10	7	5
12	17	13	9	7
13	21	17	12	9
14	25	21	15	12
15	30	25	19	15
16	35	29	23	19
17	41	34	27	23
18	47	40	32	27
19	53	46	37	32
20	60	52	43	37

▲ **Figure 4.1** Wilcoxon signed-rank test, critical vales of T

The test you are conducting is one-tailed because you are trying to decide whether your data indicate that the median has increased; that is, whether the sum of ranks of values greater than 270.5 is significantly larger than the sum of ranks of values less than 270.5. The table is constructed to give the largest value of the rank sum that can be regarded as significantly *smaller* than chance would suggest, so it is the sum of ranks corresponding to the smaller rank-sum (those ratings below 270.5) that provide the test statistic; it is often denoted by T.

You have a sample size of 15, and for a one-tailed test at the 5% significance level the table gives a critical value of 30, so this is the number to be compared with the test statistic. This means that any value less than or equal to 30 for the sum of the ranks of the ratings below 270.5 lies in the critical region. In this example, the data give a test statistic $T = 15$, so you can reject the null hypothesis in favour of the alternative hypothesis.

There is sufficient evidence to suggest that the median number of house martins is greater than 270.5.

Formal procedure for the Wilcoxon signed-rank test

The work in the previous example may be stated more formally.

» The hypotheses to be tested are as follows.

H_0: The population median of a random variable is equal to a given value M.

H_1: Either (a) the population median $\neq M$ ← two-tailed test

or (b) the population median $> M$ ←

or (c) the population median $< M$. ← one-tailed test

» The null hypothesis is based on the assumption that the random variable is symmetrically distributed about its median.

» The data are the values $x_1, x_2, \ldots, x_n$ of the random variable from a sample of size n.

» To calculate the test statistic you take the following steps.

1 Calculate the absolute differences between each sample value and the hypothesised median, M, i.e.

$$x_1 - M, x_2 - M, \ldots, x_n - M.$$

2 Rank these values from 1 to n, giving the lowest rank to the smallest absolute difference.

3 Calculate the sum W_+ of the ranks of the sample values that are greater than M, and the sum W_- of the ranks of the sample values that are less than M.

4 Check that $W_+ + W_- = \frac{1}{2} n(n + 1)$; this must work because the right-hand side is the formula for the sum of the numbers from 1 to n, that is, the total of all the ranks.

5 The test statistic T is then found as follows.

(a) For the two-tailed alternative hypothesis, take the test statistic T to be the smaller of W_- and W_+.

(b) For the one-tailed alternative hypothesis, take $T = W_-$.

(c) For the one-tailed alternative hypothesis, take $T = W_+$.

6 Reject the null hypothesis if T is *less than or equal to* the appropriate critical value found in the tables, which depends on the sample size, the chosen significance level and whether the test is one-tailed or two-tailed.

> **Note**
>
> If the assumption of symmetry of the distribution about its median is true, then this median is also the mean.

> › This test is sometimes called the Wilcoxon single-sample test and at other times the Wilcoxon signed-rank test. Explain these two names. ?

> **Note**
>
> Often the test is simply referred to as the Wilcoxon test.

Rationale for the Wilcoxon test

As you have seen, the sum of W_+ and W_- is determined by the sample size, so that the criterion for rejecting the null hypothesis is that the difference between W_+ and W_- is large enough. What makes this difference large?

Answers to exercises are available at www.hoddereducation.com/cambridgeextras

Take the case where W_+ is much larger than W_-. (The rationale is just the same if it is W_- that is much larger.) This requires W_+ to contain *more* ranks, or *larger* ranks than W_-. This will occur if *most* of the sample or the *more extreme* values (those furthest from the hypothesised median) in the sample are above the median rather than below. The largest difference between W_+ and W_- will therefore occur if the sample contains only a few values just below the hypothesised median and many values well above the hypothesised median. This is exactly the situation that would cast most doubt on the claim that the suggested median is the true one.

Example 4.4

It is known from previous trials that with a fully charged battery, a vacuum cleaner will be able to clean for a median time of 27.35 minutes before the battery runs out. The manufacturer is testing a new type of battery to see whether the cleaning time is different. The cleaning times for a random sample of 9 occasions with the new battery (rounded to 1 decimal place) are as follows.

You may remember that this is the same data as you used in Example 4.2 to carry out a sign test.

→ 25.4   28.3   26.7   29.2   30.6   28.7   27.2   27.6   29.6

Carry out a Wilcoxon test to examine, at the 5% significance level, whether there has been a change in the average cleaning time.

Solution

H_0: The median cleaning time is equal to 27.35 minutes.

H_1: The median cleaning time is different from 27.35 minutes.

Significance level: 5%

Two-tailed test.

The table shows the data, together with differences from the median and the ranks.

Time	Time − 27.35	|Time − 27.35|	Rank
25.4	−1.95	1.95	7
28.3	0.95	0.95	4
26.7	−0.65	0.65	3
29.2	1.85	1.85	6
30.6	3.25	3.25	9
28.7	1.35	1.35	5
27.2	−0.15	0.15	1
27.6	0.25	0.25	2
29.6	2.25	2.25	8

The sum of the ranks for the ratings below the median of 27.35 is

$$W_- = 7 + 3 + 1 = 11$$

and the sum of the ranks for the ratings above the median of 27.35 is

$$W_+ = 4 + 6 + 9 + 5 + 2 + 8 = 34$$

Check: $W_- + W_+ = 11 + 34 = 45 = \frac{1}{2} \times 9 \times 10$

This is a two-tailed test, so the test statistic, T, is taken to be the smaller of W_+ and W_-, which in this case is $W_- = 11$.

From the tables, the critical value for a two-tailed test on a sample of size 9, using the 5% significance level, is 5. But $11 > 5$, so you accept the null hypothesis.

There is no significant evidence that the there has been a change in the average cleaning time.

Normal approximation

The tables only give critical values where the sample size, n, is at most 20. For larger sample sizes, you use the fact that, under the null hypothesis, the test statistic T is approximately normally distributed with mean $\frac{n(n+1)}{4}$ and variance $\frac{n(n+1)(2n+1)}{24}$. Since T is discrete, you need to use a continuity correction in calculating probabilities.

Example 4.5

The median weight of yam roots of a particular variety using the standard fertiliser is 5.6 kg. The weights in kg of a random sample of 30 roots of this variety grown using a new type of fertiliser are given below. Carry out a Wilcoxon test at the 10% significance level to investigate whether roots grown with this fertiliser are on average heavier than those grown with the usual one.

5.32	4.47	7.30	7.09	6.85	4.68	6.41	8.21	3.57	5.46
5.96	5.72	6.18	5.99	6.54	4.78	6.11	5.22	7.21	4.67
5.81	7.07	5.44	6.69	5.71	7.18	5.20	6.33	5.27	5.73

Solution

The hypotheses for the test are:

H_0: The median weight with the new fertiliser is equal to 5.6.

H_1: The median weight with the new fertiliser is more than 5.6.

Weight − 5.6

−0.28	−1.13	1.70	1.49	1.25	−0.92	0.81	2.61	−2.03	−0.14
0.36	0.12	0.58	0.39	0.94	−0.82	0.51	−0.38	1.61	−0.93
0.21	1.47	−0.16	1.09	0.11	1.58	−0.40	0.73	−0.33	0.13

➙

Answers to exercises are available at www.hoddereducation.com/cambridgeextras

Ranking

7	22	28	25	23	18	16	30	29	4
9	2	14	11	20	17	13	10	27	19
6	24	5	21	1	26	12	15	8	3

The sum of the rankings for weights below the median of 5.6 is

$$W_- = 7 + 22 + 18 + 29 + 4 + 17 + 10 + 19 + 5 + 12 + 8 = 151$$

and the sum of the rankings for weights above the median of 5.6 is

$$W_+ = 28 + 25 + 23 + 16 + 30 + 9 + 2 + 14 + 11 + 20 + 13 + 27 + 6$$
$$+ 24 + 21 + 1 + 26 + 15 + 3 = 314$$

Check: $W_+ + W_- = 151 + 314 = 465 = \frac{1}{2} \times 30 \times 31$

The test statistic $T = 151$. Because the value of n is greater than 20, you cannot use the Wilcoxon tables. Instead you can use the normal approximation, which states that for large values of n the test statistic T is approximately normally distributed with mean $\frac{n(n+1)}{4}$ and variance $\frac{n(n+1)(2n+1)}{24}$.

So, in this case, the distribution of the test statistic is approximately $N(232.5, 2363.75)$.

You need to find $P(T \leqslant 151)$ (using a continuity correction) so your transformed test statistic $T' = \frac{151.5 - 232.5}{\sqrt{2363.75}} = -1.667$.

The critical value at the 10% significance level (one-tailed) $= -1.282$.

$|-1.667| > |-1.282|$ so there is sufficient evidence to reject H_0.

There is sufficient evidence to suggest that the median weight with the new fertiliser is more than 5.6 kg.

Example 4.6

Determine the 1% one-tailed (2% two-tailed) critical value for a sample of size 84.

Solution

You want to find the integer t so that

$$P(T \leqslant t) \leqslant 0.01$$

where T has mean $\frac{84 \times 85}{4} = 1785$ and variance $\frac{84 \times 85 \times 169}{24} = 50277.5$.

Note that T is a discrete variable, so that you must make a continuity correction

$$P(T \leqslant t) \approx P(\text{normal approximation to } T < t + 0.5).$$

Thus you require

$$\Phi\left(\frac{t + 0.5 - 1785}{\sqrt{50277.5}}\right) \leqslant 0.01$$

$$\text{so } t \leqslant 1784.5 + \sqrt{50277.5}\ \Phi^{-1}(0.01)$$
$$= 1784.5 - \sqrt{50277.5}\ \Phi^{-1}(0.99)$$
$$= 1784.5 - \sqrt{50277.5} \times 2.326$$
$$= 1262.95$$

This means that values of T less than or equal to 1262 are in the critical region; 1262 is the critical value.

Why Wilcoxon?

Both the Wilcoxon test and the t-test are testing whether the distribution of a random variable in a population has a given value of a **location parameter**. A location parameter is any parameter that, when it varies, shifts the position of all the values taken by the random variable but not the shape of the distribution. For instance, in the family of normal distributions, the mean is a location parameter but the variance is not; in the family of uniform distributions, the mid-range is a location parameter but the range is not.

> You will meet the uniform distribution in Chapter 5.

The value of the Wilcoxon test is that it does not make the rather strict distributional assumption of the t-test that the distribution of the random variable is normal. It is therefore very useful when this assumption is not thought to be justified, and when the sample size is not large enough for the sample means, nevertheless, to be normally distributed.

In fact, although the Wilcoxon test places a less severe restriction than the t-test on the family of distributions that the underlying variable might possess, it is nonetheless of comparable power when compared to the t-test under a wide range of conditions. This means that it is a sensible choice for testing location, even when a t-test might also be justifiable.

4.2 Paired-sample non-parametric tests

You may recall using the t-distribution for paired samples in Chapter 2. Both of the single-sample tests above (the sign test and the Wilcoxon signed-rank test) can also be also used when you have a paired sample. The general procedure is to find the differences between the first and second value for each pair. You then carry out a single-sample test on these differences.

Answers to exercises are available at www.hoddereducation.com/cambridgeextras

4

Example 4.7

A few years ago, some people believed that maths and English High School exams were of different levels of difficulty and it was claimed that candidates were getting, on average, at least one and a quarter grades higher in English than in maths. A group of eight students who took their High School exams at that time gained the following results in maths and English.

Student	1	2	3	4	5	6	7	8
Maths grade	A	D	E	C	C	E	B	E
English grade	B	C	B	A	E	A	E	E

(i)   Make a table showing how many grades better each student was in English than in maths.

(ii)   State null and alternative hypotheses for a test to investigate whether there is a difference of as much as one and a quarter grades between maths and English results.

(iii) Use the data to carry out a paired–sample sign test at the 5% significance level.

(iv)   Use the data to carry out a Wilcoxon matched-pairs signed-rank test at the 5% significance level.

> To carry out either the sign test or the Wilcoxon test, you simply use the difference in grades as a single variable.

Solution

(i)

Student	1	2	3	4	5	6	7	8
Grades better in English	−1	1	3	2	−2	4	−3	0

(ii) H_0: The median number of grades better in English is 1.25.

H_1: The median number of grades better in English is less than 1.25.

(iii) You can carry out the sign test as follows.

If the null hypothesis is true, then the probability that the 'grades better' is greater than 1.25 is $\frac{1}{2}$; similarly, the probability that it is less than 1.25 is also $\frac{1}{2}$. The number less than 1.25, X, will therefore have a binomial distribution $B(8, \frac{1}{2})$.

$$P(X = 8) = 0.0039$$

$$P(X = 7) = 0.0313$$

$$P(X = 6) = 0.1094$$

So $P(X \geqslant 7) = 0.0039 + 0.0313 = 0.0352 \leqslant 0.05$

but $P(X \geqslant 6) = 0.0039 + 0.0313 + 0.1094 = 0.1445 > 0.05$

> As in Example 4.2, adding the individual probabilities gives $P(X \geqslant 6) = 0.1446$, but working to more decimal places and then rounding gives 0.1445, which is correct to 4 d.p.

For H_0 to be rejected, at least 7 out of the 8 students would have had a difference of grades of less than 1.25.

So the critical region is $[7, 8]$.

On 5 of the 8 trials, the number of grades better was less than 1.25.

5 does not lie in the critical region.

So there is insufficient evidence to suggest that the median number of grades better in English is less than 1.25.

(iv) You can instead carry out a Wilcoxon test as follows (again using a 5% significance level).

The number of grades better in English is the value, g, of the random variable, G, which you are assuming is symmetrically distributed about its hypothesised median.

To carry out the Wilcoxon test, you use exactly the same procedure as for a single-sample Wilcoxon test, but you use the difference in values (in this case, the number of grades better) as the variable.

Student	g	$g - 1.25$	$|g - 1.25|$	Rank
1	−1	−2.25	2.25	5
2	1	−0.25	0.25	1
3	3	1.75	1.75	4
4	2	0.75	0.75	2
5	−2	−3.25	3.25	7
6	4	2.75	2.75	6
7	−3	−4.25	4.25	8
8	0	−1.25	1.25	3

So that: $W_+ = 4 + 2 + 6 = 12$

$\qquad W_- = 5 + 1 + 7 + 8 + 3 = 24.$

Check: $W_+ + W_- = 12 + 24 = 36 = \frac{1}{2} \times 8 \times 9$ as required.

The test statistic is $T = W_+ = 12$.

From the tables, using the one-tailed 5% significance level, the critical value for a sample size of 8 is 5. Since $12 > 5$ you accept the null hypothesis.

There is insufficient evidence to suggest that the median number of grades better in English is less than 1.25.

General method for a Wilcoxon matched-pairs signed-rank test

In general, when the values of a random variable have been measured on a sample in two different conditions, you may want to test the hypothesis that the medians differ by some given amount between the two conditions.

In this situation the Wilcoxon matched-pairs signed-rank test procedure is as follows.

» Find the differences between the values in the two conditions.

» Use the single-sample Wilcoxon signed-rank test with the hypothesis that these differences have the suggested median.

Answers to exercises are available at www.hoddereducation.com/cambridgeextras

The distributional assumption is then that, in the population, the differences between the values of the random variables in the two conditions are symmetrically distributed about the median difference.

Perhaps the most natural context in which this test is used is when you are trying to detect a shift in a location parameter of the distribution of a random variable between two conditions.

It is worth noticing that the Wilcoxon matched–pairs signed-rank test and the single-sample Wilcoxon signed-rank test are related in the same way as the paired-sample *t*-test and the single-sample *t*-test.

Example 4.8	

Seven randomly selected economists were asked on two occasions to predict what the growth rate of GDP would be for the year ending in December 2015 in a particular country. Their predictions, made in June 2014 and December 2014 were as listed below.

Economist	A	B	C	D	E	F	G
June 2014 prediction (%)	2.2	3.7	2.1	2.3	3.4	2.5	2.1
December 2014 prediction (%)	2.9	3.8	2.6	3.1	3.0	2.7	2.7

Use a Wilcoxon test to examine whether there is evidence at the 5% level that economists became more optimistic between June and December about the future growth rate.

Solution

The increases in the predictions between June and December 2014 are as shown.

Economist	A	B	C	D	E	F	G
Increase in prediction (%)	0.7	0.1	0.5	0.8	−0.4	0.2	0.6

The hypotheses under test are:

H_0: The median increase in prediction is zero.

H_1: The median increase in prediction is positive.

So it is the absolute increases in prediction that you need to rank.

Economist	A	B	C	D	E	F	G
Increase in prediction (%)	0.7	0.1	0.5	0.8	−0.4	0.2	0.6
Absolute increase in prediction (%)	0.7	0.1	0.5	0.8	0.4	0.2	0.6
Rank	6	1	4	7	3	2	5

Hence

$$W_+ = 6 + 1 + 4 + 7 + 2 + 5 = 25$$
$$W_- = 3$$

Check: $W_+ + W_- = 25 + 3 = 28 = \frac{1}{2} \times 7 \times 8$

The test statistic is therefore: $T = W_- = 3$.

From the tables, the critical value at the 5% level for a sample of size 7 is 3 and $3 \leqslant 3$.

So the test statistic lies in the critical region and therefore you reject the null hypothesis in favour of the alternative that economists have become more optimistic.

Exercise 4A

1 At a bird observatory in West Africa, the numbers of black-headed weavers flying past, recorded for a sample of 12 days this year, are as follows.

    134   162   108   144   165   184   147   151   131   98   126   148

Records show that, for several years, the median number per day has been 120. Use a sign test to examine, at the 5% significance level, whether there has been an increase in the numbers of black-headed weavers flying past the observatory this year.

2 A manager is monitoring the number of weeks that patients wait to get an appointment after being referred to a dermatology clinic. She has found the median waiting time is 29 weeks. A new procedure for referrals is introduced and the manager wishes to find out whether the average waiting time has decreased.

Following the introduction of the new procedure, the waiting times of a random sample of 14 patients are as follows.

    33   26   28   21   26   32   30   27   25   19   22   33   30

Use a sign test to examine, at the 5% significance level, whether there has been a decrease in the average waiting time.

3 A racing driver wishes to decrease her average lap time on a particular circuit from its present value of 184 seconds. Her mechanic makes an adjustment to the brakes in order to achieve this. After this adjustment, the mechanic records the lap times for a sample of 60 laps. Of these, 49 take less than 184 seconds. Use a sign test to examine, at the 5% significance level, whether there has been a decrease in the average lap time.

4 A two-tailed Wilcoxon test is carried out on a sample of size 15. The values of W_- and W_+ are 84 and 36, respectively.

   (i)   Write down the critical value for the test at the 5% level.

   (ii)   State whether the null hypothesis should be accepted or rejected.

Answers to exercises are available at www.hoddereducation.com/cambridgeextras

5 A Wilcoxon test is carried out to investigate the following hypotheses.

H_0: The ratings are symmetrically distributed about a median of 25.

H_1: The ratings are symmetrically distributed about a median greater than 25.

The values of W_- and W_+ are 8 and 37, respectively.

(i) Find the sample size (assuming that none of the values in the sample is equal to 25).

(ii) Write down the critical value for the test at the 5% level.

(iii) Complete the test.

Ⓜ 6 Gerry runs 5000 m races for his local athletics club. His coach has been monitoring his practice times for several months and he believes that their median is 15.3 minutes. The coach suggests that Gerry should try running with a pacemaker in order to see if this can improve his times. Subsequently, a random sample of ten of Gerry's times with the pacemaker is collected to see if any reduction has been achieved. A test is carried out using the hypotheses below to investigate whether the times have reduced.

H_0: The times are symmetrically distributed about a median of 15.3.

H_1: The times are symmetrically distributed about a median less than 15.3.

The spreadsheet output below shows the calculations for a Wilcoxon test to investigate this.

	A	B	C	D	E	F
	Time	**Time – 15.3**	**[Time – 15.3]**	**Rank**	**Negative**	**Positive**
1						
2	14.86	−0.44	0.44	7	7	
3	15.00	−0.30	0.30	4	4	
4	15.62	0.32	0.32	5		5
5	14.44	−0.86	0.86	9	9	
6	15.27	−0.03	0.03	1	1	
7	15.64	0.34	0.34	6		6
8	14.58	−0.72	0.72	8	8	
9	14.30	−1.00	1.00	10	10	
10	15.08	−0.22	0.22	2	2	
11	15.07	−0.23	0.23	3	3	
12				**Sum**	**44**	**11**

(i) Write down the values of W_- and W_+.

(ii) Explain why the sum of W_- and W_+ must be equal to 55.

(iii) Carry out the test at the 5% level of significance.

Ⓜ 7 An ancient human settlement site in the Harz mountains has been explored by archaeologists over a long period. They have established by a radiocarbon method that the ages of bones found at the site are approximately uniformly distributed between 3250 and 3100 years. A new potassium–argon method of dating has now been developed and 11 samples of bone randomly selected from finds at the site are dated by this new method. The ages, in years, determined by the new method are as listed below.

3115	3234	3247	3198	3177	3227
3124	3204	3166	3194	3220	

Is there evidence at the 5% level that the potassium–argon method is producing different dates, on average, for bones from the site?

8 A local education authority sets a reasoning test to all eleven-year-olds in the area. The scores of the whole area on this test have been symmetrically distributed around a median of 24 out of 40 over many years.

One year a random sample of 11 of a particular primary school's leavers have the following scores out of 40.

34	26	23	40	31	18	21	37	35	33	20

Is there evidence at the 5% level to support the headteacher's claim that her leavers score better on the reasoning test than average?

When must this claim have been made if the hypothesis test is to be valid?

9 Becotide inhalers for people with asthma are supposed to deliver 50 mg of the active ingredient per puff. In a test in a government laboratory, 17 puffs from randomly selected inhalers in a batch were tested and the amount of active ingredient that was delivered was determined. The results, in milligrams, are given below.

43.9	47.1	52.3	51.4	45.0	50.6	51.3	41.0	49.0
46.1	52.6	50.5	47.8	45.2	49.9	46.0	42.8	

Is there evidence at the 2% level that the inhalers are not delivering the correct amount of active ingredient per puff?

10 When a consignment of grain arrives at Rotterdam docks, the percentage of moisture in 11 samples is measured. It is claimed that when the ship left Ontario, the percentage of moisture in the grain was 2.353%, on average.

The percentages found in the samples were as follows.

5.294	0.824	3.353	1.706	3.765	3.235
8.235	0.760	3.412	6.471	3.471	

(i) Test at the 5% level whether the median percentage of moisture in the grain is greater than 2.353, using the Wilcoxon single-sample test. What assumptions are you making about the distribution of the percentage of moisture in the grain?

(ii) Test at the 5% level whether the mean percentage of moisture in the grain is greater than 2.353, using the *t*-test. What assumption are you making about the distribution of the percentage of moisture in the grain?

(iii) Compare your two conclusions and comment.

Answers to exercises are available at www.hoddereducation.com/cambridgeextras

4

PS **11** Check the claim in the tables that the critical value for the Wilcoxon single-sample test, at the 5% level, for a sample of size nine is 8.
Tip: There are $2^9 = 512$ different sets of ranks that, under the null hypothesis, are equally likely to make up W_+ but it is only necessary to write down those with the smallest rank sums − the sets giving rank sums up to and including 9 are sufficient to verify the result. For example, the sets $\{1\}, \{2, 3\}, \{1, 3, 5\}$ all have sums of 9 or less.

M **12** A company is developing a new small scale wind turbine to convert wind energy into electricity. Current models produce, on average, 500 units of electricity per day. The new model is tested over a period of several months. The number of units of electricity generated per day on 50 particular days is recorded. For a Wilcoxon test, the sum of the ranks of those days where more than 500 units of electricity is generated is 775. Stating any necessary assumptions, use a Wilcoxon test to examine, at the 10% significance level, whether the new model produces more energy, on average, than current models.

M **13** The amount of nitrogen dioxide in airborne dust (measured in suitable units) at 20 sampling spots around a large city was measured before and after government measures to encourage the use of less-polluting vehicles were introduced.

Examine whether the data give evidence at the 2.5% level that the government measures have reduced the average amount of nitrogen dioxide in the air

(i) using a sign test

(ii) using a Wilcoxon test.

Amount of nitrogen dioxide		Amount of nitrogen dioxide	
Before	After	Before	After
43	47	23	56
11	8	29	17
133	102	78	57
28	34	170	138
91	72	61	56
48	41	14	13
89	91	40	27
205	196	167	157
81	15	80	97
111	103	19	8

M **14** The speed with which 13 subjects react to a stimulus is timed in hundredths of a second. They then play a computer game for 20 minutes and their reaction times are re-measured.

Subject	Reaction time	
	Before game	After game
A	231	201
B	337	344
C	168	183
D	243	215
E	197	188
F	205	181
G	265	291
H	170	175
I	302	281
J	250	242
K	316	306
L	252	211
M	226	193

Examine whether there is evidence at the 5% level that playing the computer game has reduced their reaction times

(i)   using a sign test   (ii)   using a Wilcoxon test.

15 A highly skilled typist is comparing the abilities of two equation editing systems for the typing of complicated mathematical expressions. Although both systems provide special facilities, they both have some difficulties in dealing with such expressions. The typist has taken several mathematical articles that have been professionally typeset and has typed each one using both equation editing systems. For each article, she has counted the number of such expressions with which, in her opinion, there has been unusual difficulty. The results are as follows.

Article	A	B	C	D	E	F	G	H	I	J
System I	6	5	11	16	5	3	12	16	17	4
System II	5	10	19	12	15	5	5	27	30	19

(i)   Carry out a Wilcoxon test, at the 5% level of significance, to examine whether the two systems have, on the whole, the same ability in coping with such expressions.

The typist notes that the test statistic is only slightly higher than the critical value, so she decides to take a much larger sample. She types 90 articles and again counts the number of expressions with which there has been unusual difficulty using each of the systems. Of these 90 articles, she finds that in 55 of them there were a larger number of difficult expressions using system 1 and 35 using system 2.

(ii)   Carry out a sign test at the 5% level of significance to investigate whether the two equation editing systems have, on the whole, the same ability in coping with such expressions.

Answers to exercises are available at www.hoddereducation.com/cambridgeextras

4.3 The Wilcoxon rank-sum test

In 2016, nine economists made predictions for the growth rate in Indian gross domestic product (GDP) for 2017. Their predictions are shown below, together with seven predictions for the growth rate in the Chinese GDP made at the same time.

Predictions for 2017 growth rate								
India	6.2	7.0	6.6	5.2	6.0	6.5	8.0	7.1	5.8
China	6.9	7.2	7.4	7.6	7.5	6.8	8.1		

Is there evidence that the economists are more optimistic about the Chinese growth rate than that of India?

This is exactly the type of question answered in Chapter 2, except that here there is really no reason to assume that the economists' predictions for the growth rate would be normally distributed. This means that you would not be justified in using a 2-sample t-test to compare their predictions. Earlier in this chapter, you saw that when the assumptions required for tests based on the normal distribution are not appropriate, it is possible to use ranking methods to compare the median performance of two groups. There a paired design was used, but the data here are unpaired. You will now develop a method, based on ranks, for comparing the medians of two populations that is analogous to the unpaired t-test in its use of two independent samples.

You can get a feeling for these data visually by ranking all 16 predictions, giving something akin to a back-to-back stem-and-leaf diagram as shown on the right.

This makes it look as though the Chinese predictions are higher on the whole; you can set this up as a hypothesis test.

India	Rank	China
5.2	1	
5.8	2	
6.0	3	
6.2	4	
6.5	5	
6.6	6	
	7	6.8
	8	6.9
7.0	9	
7.1	10	
	11	7.2
	12	7.4
	13	7.5
	14	7.6
8.0	15	
	16	8.1

| Example 4.9 | Use the data given on the previous page to test, at the 5% significance level, whether there is evidence that the economists are more optimistic about the Chinese growth rate than that of India. |

Solution

The formal statement is:

H_0: The population median predictions for India and China are equal.

H_1: The population median prediction for China is greater than that for India.

Significance level: 5%

One-tailed test.

In order to carry out any hypothesis test, the sample(s) must be random. We are interested in population rather than sample medians.

If you add the ranks of the Indian and Chinese predictions separately, you get the Wilcoxon statistics.

$$W_I = 1 + 2 + 3 + 4 + 5 + 6 + 9 + 10 + 15 = 55$$
$$W_C = 7 + 8 + 11 + 12 + 13 + 14 + 16 = 81$$

Check: $W_I + W_C = 55 + 81 = 136$

This total should equal the sum of all the ranks,

$$1 + 2 + \dots + 16 = \frac{1}{2} \times 16 \times 17$$
$$= 136$$

which it does.

You use the value for the smaller set to find the test statistic W. In this case, the Chinese prediction with a value of 81 is the value from the smaller set. However, as you are interested in the upper tail, you must subtract this value of 81 from $m \times (n + m + 1) = 7 \times (9 + 7 + 1) = 119$. This gives the test statistic $W = 119 - 81 = 38$.

The critical value is found using the table of critical values for the Wilcoxon rank-sum test for the relevant sample sizes and significance level (see Figure 4.2). In this case, the sample sizes are 7 and 9, the significance level is 5% and the test is one-tailed, so the critical value is 43.

$38 < 43$, so the test statistic of 38 lies within the critical region.

The null hypothesis is therefore rejected in favour of the alternative hypothesis that the population median prediction for China is greater than that for India.

Figure 4.2 on the next page shows an extract from the table of critical values of the Wilcoxon rank-sum test.

The two samples have sizes m and n, where $m \leqslant n$.

R_m is the sum of the ranks of the items in the sample of size m.

W is the smaller of R_m and $m(n + m + 1) - R_m$.

For each pair of values of m and n, the table gives the *largest* value of W that will lead to rejection of the null hypothesis at the level of significance indicated.

Answers to exercises are available at www.hoddereducation.com/cambridgeextras

	Level of significance											
One-tailed	0.05	0.025	0.01	0.05	0.025	0.01	0.05	0.025	0.01	0.05	0.025	0.01
Two-tailed	0.1	0.05	0.02	0.1	0.05	0.02	0.1	0.05	0.02	0.1	0.05	0.02
n	$m = 3$			$m = 4$			$m = 5$			$m = 6$		
3	6	–	–									
4	6	–	–	11	10	–						
5	7	6	–	12	11	10	19	17	16			
6	8	7	–	13	12	11	20	18	17	28	26	24
7	8	7	6	14	13	11	21	20	18	29	27	25
8	9	8	6	15	14	12	23	21	19	31	29	27
9	10	8	7	16	14	13	24	22	20	33	31	28
10	10	9	7	17	15	13	26	23	21	35	32	29

	Level of significance											
One-tailed	0.05	0.025	0.01	0.05	0.025	0.01	0.05	0.025	0.01	0.05	0.025	0.01
Two-tailed	0.1	0.05	0.02	0.1	0.05	0.02	0.1	0.05	0.02	0.1	0.05	0.02
n	$m = 7$			$m = 8$			$m = 9$			$m = 10$		
7	39	36	34									
8	41	38	35	51	49	45						
9	43	40	37	54	51	47	66	62	59			
10	45	42	39	56	53	49	69	65	61	82	78	74

For larger values of m and n, the normal distribution with mean $\frac{1}{2}m(m + n + 1)$ and variance $\frac{1}{12}mn(m + n + 1)$ should be used as an approximation to the distribution of R_m.

▲ **Figure 4.2** Wilcoxon rank-sum test, critical values of W

Note

If you had been interested in the lower tail, you would have simply compared the value of W_c with the critical value of 43. It is because you are dealing with the upper tail that you have to subtract from $m \times (n + m + 1)$.

There is an alternative method of performing the test. This compares the value of $W_c = 81$ to the 'upper tail critical value'. In order to find this, it is important to understand the range of possible values for W_c and the relationship of the critical regions to it.

In this example, the smaller sample has size 7 out of a total of 16.

The smallest possible value of W_c corresponds to ranks 1, 2, 3, ..., 7, totalling $\frac{1}{2} \times 7 \times 8 = 28$.

The largest possible value of W_c corresponds to ranks 10, 11, 12, 13, 14, 15, 16. In this case $\frac{1}{2} \times 16 \times 17 - \frac{1}{2} \times 9 \times 10 = 91$.

The mean value is $\frac{1}{2} \times (28 + 91) = 59.5$.

The critical value of 43 given in the tables corresponds to the lower tail. It is $(59.5 - 43) = 16.5$ below the mean. The equivalent value for the upper tail is 16.5 above the mean: $59.5 + 16.5 = 76$. You could have compared the value of W_c of 81, which you found above, to this upper tail critical value. So you could have made the comparison $81 > 76$ and come to the same conclusion as when you compared 38 to 43. You will note that the difference in each case is 5.

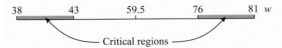

▲ Figure 4.3

If you were carrying out a two-tailed test at the 10% significance level, you could use the lower critical value of 43 and the upper tail critical value of 76.

The Wilcoxon rank-sum test is appropriate for tests of hypotheses similar to those for which the 2-sample (unpaired) *t*-tests that you met in Chapter 2 were used. Note the absence of the special cases that you had to consider there for small or large samples, equal or unequal variances, normal or non-normal underlying distributions. This *non-parametric* test is applicable in the same single form to a wide variety of situations where *difference of location* in two conditions is being tested.

Note

You will note that in the table of predictions for India and China, none of the figures are equal to each other. If any had been equal you would have had to give the average of the two ranks, which would otherwise have been given. For example, if two figures for India were both equal to 6.5 (which would have ranks 5 and 6 if they were almost, but not quite, equal), you would give them both a **tied rank** of 5.5.

Tied ranks are an extension beyond the requirements of the Cambridge International syllabus.

Answers to exercises are available at www.hoddereducation.com/cambridgeextras

Formal procedure for the Wilcoxon rank-sum test

To calculate the statistic you take the following steps.

1 Rank all $(m + n)$ sample values 1 to $(m + n)$, giving the lowest rank to the smallest value.

2 Calculate the sum, W_X, of the ranks of the x-values, and the sum, W_Y, of the ranks of the y-values.

3 Check that $W_X + W_Y = \frac{1}{2}(m + n)(m + n + 1)$.

This must work because the right-hand side is the formula for the sum of the numbers from 1 to $(m + n)$; that is, the total of all the ranks.

4 For a lower tail test, the test statistic is the value of W for the smaller sample, that of size m. For an upper tail test, the test statistic is $m(m + n + 1) - W$. For a two-tailed test, the test statistic is the smaller of W and $m(m + n + 1) - W$.

5 The critical value is found from the table of critical values for the Wilcoxon rank-sum test, according to the two sample sizes, the chosen significance level and whether the test is one- or two-tailed.

6 The null hypothesis is then rejected if W is at least as extreme as the appropriate critical value (in other words, less than the critical value).

Example 4.10

In an experiment on cultural differences in interpreting facial expressions, 10 English children and 12 Japanese children were each shown 40 full-face photographs of American children who had been asked to display particular emotions in their facial expressions.

They were asked to describe the feelings of the children in the photographs and the number of correct responses they made was noted. The results of this experiment, as scores out of 40, are shown below.

English children	16	18	19	23	28	29	34	35
	37	39						
Japanese children	17	21	24	25	26	27	30	31
	33	36	38	40				

Is there evidence at the 5% level that English and Japanese children differ in their abilities to identify American facial expressions?

Solution

The formal statement of the hypotheses under test is:

H_0: The distributions of English and Japanese scores have the same median.

H_1: The distributions of English and Japanese scores have different medians,

under the assumption that the distributions have the same shape.

Significance level: 5%

Two-tailed test.

The scores are shown below, with their ranks.

English scores	Rank	Japanese scores
16	1	
	2	17
18	3	
19	4	
	5	21
23	6	
	7	24
	8	25
	9	26
	10	27
28	11	
29	12	
	13	30
	14	31
	15	33
34	16	
35	17	
	18	36
37	19	
	20	38
39	21	
	22	40

The sum of the rankings for English scores is

$$W_E = 1 + 3 + 4 + 6 + 11 + 12 + 16 + 17 + 19 + 21 = 110$$

and the sum of the rankings for Japanese scores is

$$W_J = 2 + 5 + 7 + 8 + 9 + 10 + 13 + 14 + 15 + 18 + 20 + 22 = 143$$

Check: $W_E + W_J = 110 + 143 = 253 = \frac{1}{2} \times 22 \times 23$

So the value of W for the smaller sample is 110. The test statistic is the lower of $W = 110$ and $m(m + n + 1) - W = 10 \times 23 - 110 = 120$. Thus the test statistic is 110.

Because the value of n is greater than 10, the tables for the Wilcoxon rank-sum test cannot be used. Instead, you can use the normal approximation, which states that, for large values of m and n, the test statistic W is approximately normally distributed with mean $\frac{1}{2}m(m + n + 1)$ and variance $\frac{1}{12}mn(m + n + 1)$.

→

Answers to exercises are available at www.hoddereducation.com/cambridgeextras

So in this case where $m = 10$ and $n = 12$

the mean is $\frac{1}{2}m(m + n + 1) = \frac{1}{2} \times 10 \times (10 + 12 + 1) = 115$

and the variance is $\frac{1}{12}mn(m + n + 1) = \frac{1}{12} \times 10 \times 12 \times (10 + 12 + 1) = 230$

so the distribution of the test statistic is approximately $N(115, 230)$.

You need to find $P(T \leqslant 110)$ so, applying a continuity correction, your

transformed test statistic $T' = \dfrac{110.5 - 115}{\sqrt{230}} = -0.2967$

> You need to apply a continuity correction since the ranks are all whole numbers.

The critical value at the 5% significance level (two-tailed) $= -1.96$.

$-0.2967 > -1.96$ so there is not enough evidence to reject H_0.

There is insufficient evidence to suggest that the English and Japanese children differ in their abilities to identify American facial expressions.

| Example 4.11 | Determine the 1% one-tailed (2% two-tailed) critical value for the samples with sizes $m = 33$ and $n = 58$. |

Solution

You want to find the integer t so that $P(T \leqslant t) \leqslant 0.01$, where T has

mean $\frac{1}{2}m(m + n + 1) = \frac{1}{2} \times 33 \times (33 + 58 + 1) = 1518$

and variance $\frac{1}{12}mn(m + n + 1) = \frac{1}{12} \times 33 \times 58 \times (33 + 58 + 1) = 14674$.

T is a discrete variable, so you must make a continuity correction.

$P(T \leqslant t) \approx P(\text{normal approximation to } T < t + 0.5)$.

Thus you require $\Phi\left(\dfrac{t + 0.5 - 1518}{\sqrt{14674}}\right) \leqslant 0.01$

> From the table of critical values for the normal distribution,
> $P(Z \leqslant 2.326) = 0.99$,
> so $P(Z \leqslant -2.326) = 0.01$

$\left(\dfrac{t + 0.5 - 1518}{\sqrt{14674}}\right) = -2.326$

$t = 1517.5 - \sqrt{14674} \times 2.326$

$= 1235.74$

This means that values of T less than or equal to 1235 are in the critical region. 1235 is the critical value.

| Exercise 4B | This exercise contains questions about both paired and unpaired samples, to enable you to practise deciding which non-parametric procedure is appropriate. |

1   Of 17 school-leavers from one school who went on to university, 7 went to the National Science University (NSU) or National University of

Technology (NUT) and 10 to other universities. Three years later, when these 17 got jobs, their starting salaries, in $, were as listed below.

NSU and NUT students

35,000   24,060   26,540   22,810   28,400   43,300   18,024

Students from other universities

37,500   23,234   30,400   20,680   22,080   15,720   21,960

25,090   23,300   23,995

(i) Test, at the 5% level, the hypothesis that the starting salaries of NSU and NUT graduates are higher than those of graduates from other universities.

(ii) Criticise the sampling procedure adopted in collecting these data.

2 The blood cholesterol levels of 46 men and 28 women are measured. These data are shown below.

Men	621	237	92	745
	301	550	182	723
	1301	56	105	428
	209	478	119	303
	417	869	384	1058
	939	1080	1104	829
	1061	145	382	919
	813	770	312	204
	610	139	206	174
	67	258	333	1203
	407	826	810	922
	717	106		
Women	207	529	104	72
	377	482	50	620
	1003	162	94	391
	149	371	208	194
	901	205	370	871
	710	973	304	189
	683	191	233	127

[Sum of men's ranks = 1878; sum of women's ranks = 897.]

Is there evidence at the 5% level that the blood cholesterol levels of men and women differ?

Do the data suggest that the assumptions of the test are justified?

Answers to exercises are available at www.hoddereducation.com/cambridgeextras

M 3 Given two rock samples, A and B, geologists say that A is harder than B if, when the two samples are rubbed together, A scratches B. Eleven rock samples from stratum X and seven from stratum Y are tested in this way to determine the order of hardness. The list below shows 18 samples in order of hardness and which stratum they come from. The hardest is on the right.

X X X X X Y X Y X X Y X Y Y Y X X Y Y

Is there evidence at the 5% level that the rock in stratum Y is harder than that in stratum X? The critical value at the 5% level for $m = 7$, $n = 11$ is 86.

M 4 An aggregate material used for road surfacing contains some small stones of various types. A highway engineer is examining the composition of this material as delivered from two separate suppliers. The stones are broadly classified into two types, rounded and non-rounded, and the percentage of the rounded type is found in each of several samples. It is desired to examine whether the two suppliers are similar in respect of this percentage.

(i) In an initial investigation, histograms are drawn of the proportions of rounded stones in samples from the two suppliers. Discuss briefly what these histograms should indicate about the shape of the underlying distribution so that the comparison may reasonably be made using

(a) t-test

(b) a Wilcoxon rank-sum test.

(ii) It is decided that a Wilcoxon rank-sum test must be used. Detailed data of the percentage of the rounded type of stone for 15 samples, 9 from one supplier and 6 from the other, are as follows.

Supplier 1	46	52	34	17	21	63	55	48	25
Supplier 2	59	53	71	39	66	58			

Test at the 5% level of significance whether it is reasonable to assume that the true median percentages for the suppliers are the same.

M 5 The canteen in a factory is some distance from the building where the actual work takes place. The company believes that its employees take longer walking back to work after breaks in the canteen than they do in walking there. They time ten people walking to the canteen and ten walking back from it. Their times, in seconds, are as follows.

Time to canteen	50	46	57	72	33	84	104	79	39	63
Time from canteen	55	73	56	88	35	109	127	75	42	76

(i) Use these data to carry out a test of the company's belief at the 5% significance level.

After the test has been carried out, the person who collected the data points out that they refer to just ten individuals. Their times are given in the same order.

(ii) Carry out a new hypothesis test in the light of this information.

(iii) Compare and comment on the results of the two tests.

6 Random samples $x_1, x_2, \ldots, x_m$ and $y_1, y_2, \ldots, y_n$ are taken from two independent populations. The Wilcoxon rank-sum test is to be used to test the following hypotheses.

H_0: The two populations have identical distributions.

H_1: The two populations have identical distributions except that their location parameters differ.

Accordingly, the complete set of $m + n$ observations is ranked in ascending order (it may be assumed that no two observations are exactly equal). S denotes the sum of the ranks corresponding to $x_1, x_2, \ldots x_m$.

(i) Consider the case $m = 4, n = 5$.

    (a) Show that, if all the x are less than all the y, then the value of S is 10.

    (b) List all possible sets of ranks of the x values that give rise to a value of S such that $S \leqslant 12$.

    (c) What is the total number of ways of assigning four ranks from the available nine to the x values?

    (d) Deduce from your answers to parts (b) and (c) that the probability that $S \leqslant 12$ if H_0 is true is $\dfrac{4}{126}$.

    (e) Compute the value of this probability as given by the normal approximation

$$N\left(\tfrac{1}{2}m(m+n+1), \tfrac{1}{12}mn(m+n+1)\right)$$

    to the distribution of S if H_0 is true.

(ii) The following are the numerical values of the data for a case with $m = 6, n = 8$.

Sample 1 $(x_1, x_2, \ldots, x_m)$	4.6	6.6	6.0	5.2	8.1	9.5		
Sample 2 $(y_1, y_2, \ldots, x_n)$	5.5	7.9	7.1	6.3	8.4	6.8	10.2	9.0

Test H_0 against H_1 at the 5% level of significance, using the normal approximation given in part (e) above to the distribution of S under H_0 or otherwise.

7 As part of the procedure for interviewing job applicants, a firm uses aptitude tests. Each applicant takes a test and receives a score. Two different tests, A and B, are used. The distribution of scores for each test, over the whole population of applicants, are understood to be similar in shape, approximately symmetrical, but not normal. However, the location parameters of these distributions may differ. The personnel manager is investigating this by considering the medians of the distributions, with the null hypothesis, H_0, that these medians are equal. It is thought

Answers to exercises are available at www.hoddereducation.com/cambridgeextras

that test B might lead to consistently lower scores, so the alternative hypothesis H_1 is that the median for test B is *less than* that for test A.

The scores from test A for a random sample of seven applicants for a particular job are as follows.

| 37.6 | 34.3 | 38.5 | 38.8 | 35.8 | 38.0 | 38.2 |

The scores from test B for a separate random sample of seven applicants for this job are as follows.

| 37.3 | 33.0 | 33.9 | 32.1 | 37.0 | 35.0 | 36.2 |

(i)   Calculate the value of the Wilcoxon rank-sum test statistic for these data.

(ii) State the critical value for a one-sided 5% test of H_0 against H_1.

(iii) State whether values *less than* this critical value would lead to the acceptance or rejection of H_0.

(iv) Carry out the test.

Now suppose, instead, that the two distributions can be taken as normal (with the same variance), and that the test is to be conducted in terms of the means. Describe briefly a procedure based on the t-distribution for carrying out this test using these data.

PS   **8**   Random samples $x_1, x_2, \ldots, x_m$ and $y_1, y_2, \ldots, y_n$ are taken from two independent populations. It is understood that these two populations have identical distributions except possibly for a difference in location parameter. The null hypothesis, H_0, that they have the same location parameter is to be tested against the alternative hypothesis, H_1, that their location parameters differ, using the Wilcoxon rank-sum test. The complete set of all $m + n$ observations is ranked in ascending order (it may be assumed that no two observations are exactly equal); W denotes the sum of the ranks corresponding to $x_1, x_2, \ldots, x_m$.

(i)   Show that the minimum value W can take is $\frac{1}{2}m(m + 1)$.

(ii) Find the maximum value W can take.

(iii) The sample data are as follows.

Sample 1 $(x_1, x_2, \ldots, x_m)$	10.2	13.5	15.2	15.4	10.8	17.7	12.1	13.8		
Sample 2 $(y_1, y_2, \ldots, x_n)$	14.6	19.3	11.6	12.9	18.4	16.9	18.2	18.7	16.1	19.0

Calculate the value of W and hence carry out a two-sided 5% test.

(iv) Show that if H_0 is true the expected value of W is $\frac{1}{2}m(m + n + 1)$. Given that the variance of W if H_0 is true is $\frac{1}{2}m(m + 1)$, repeat the test in part (iii) using a normal approximation to the distribution of W under H_0.

9 While at a summer camp, children are given healthy food and take plenty of exercise. In their publicity, the camp's organisers claim that after a stay the children are 'leaner and fitter'.

An advertising watchdog body decide to check this claim. They weigh 10 children at the start of their stay and when they leave, three weeks later. The 10 children are also required to run 1.5 km at the start and end of their stay; on both occasions, the times taken by the 1st, 2nd, …, 10th child to complete the distance are recorded. The data are given below.

Weights (kg)			Times (minutes and seconds)		
Name	**Start**	**End**	**Place**	**Start**	**End**
Wei	48	53	1	7m 55s	7m 39s
Dexter	64	58	2	7m 56s	7m 52s
Ying	49	47	3	8m 12s	8m 05s
Asha	35	38	4	8m 13s	8m 30s
Hakim	47	40	5	8m 46s	8m 55s
Vinoj	62	54	6	9m 35s	8m 56s
Nijah	38	39	7	10m 34s	10m 01s
Jie	57	47	8	11m 22s	15m 10s
Cassie	68	57	9	15m 38s	15m 11s
Qiang	38	33	10	19m 22s	15m 12s

(i) Explain why it is not appropriate to use tests based on the normal distribution or the *t*-distribution.

(ii) Use appropriate tests at the 5% significance level on these data to investigate the camp's organisers' claim that after a stay the children are 'leaner and fitter'.

(iii) Do the camp's organisers have grounds to criticise the experiment?

Answers to exercises are available at www.hoddereducation.com/cambridgeextras

KEY POINTS ✓

1   The sign test is used to test the value of the population median m or the median difference m_d between two populations when samples are paired.

No distributional assumption is required, but the magnitude of the differences is ignored.

2   The Wilcoxon signed-rank test is used to test the value of the population median m or the median difference m_d between two populations when samples are paired.

- Distributional assumption required: the variable is symmetrically distributed about its median.

- P is the sum of the ranks corresponding to the positive differences; Q is the sum of the ranks corresponding to the negative differences.

- The test statistic T is the smaller of P and Q.

- The critical value (the smallest value of T for which H_0 is rejected) is found from tables.

- If n is large, T is approximately distributed as
$N\left(\dfrac{n(n+1)}{4}, \dfrac{n(n+1)(2n+1)}{24}\right)$ and a continuity correction should be used.

3   The Wilcoxon rank-sum test is used to test the null hypothesis that two populations have the same median.

- Distributional assumption required: the distributions of the two populations have the same shape.

- The two samples have sizes m and n, where $m \leqslant n$.

- R_m is the sum of the ranks for the sample of size m.

- The test statistic W is the smaller of R_m and $m(m+n+1) - R_m$.

- The critical value (the smallest value of W for which H_0 is rejected) is found from tables.

- If m and n are large, W is approximately distributed as
$N\left(\dfrac{m(m+n+1)}{2}, \dfrac{mn(m+n+1)}{12}\right)$ and a continuity correction should be used.

LEARNING OUTCOMES

Now you have finished this chapter, you should be able to

- understand the idea of a non-parametric test and appreciate situations in which such a test might be useful
- use a single-sample sign test to test a hypothesis concerning a population median
- use a single-sample Wilcoxon signed-rank test to test a hypothesis concerning a population median
- use a paired-sample sign test to test for identity of two paired populations
- use a Wilcoxon matched-pairs signed-rank test to test for identity of two paired populations
- use a Wilcoxon rank-sum test to test for identity of two separate populations.

Answers to exercises are available at www.hoddereducation.com/cambridgeextras

Probability generating functions

During the first season of England's Premier League football competition, 462 matches were played. The frequency of the number of goals scored in each game by the *away* team, and associated relative frequencies, were recorded as follows.

Number of goals	0	1	2	3	4	5	> 5
Frequency	149	179	91	37	3	3	0
Relative frequency (to 3 d.p.)	0.323	0.387	0.197	0.080	0.006	0.006	0

Assuming that these relative frequencies reflect the general pattern of scoring in Premier League matches, they may be used as *empirical* probabilities of 0, 1, 2, … goals being scored by *away* teams per match.

5.1 Probabilities defined by a probability generating function

Let X be the discrete random variable representing the number of goals scored per game by away teams. Then one way of storing the probabilities is in the **probability generating function (PGF)**

$$G(t) = 0.323 + 0.387 \times t + 0.197 \times t^2 + 0.080 \times t^3$$
$$+ 0.006 \times t^4 + 0.006 \times t^5$$

where the coefficients of t^x are the probabilities $P(X = x)$. The variable t is a **dummy** variable as, in itself, it has no significance. However, it has an important role to play in subsequent analysis, which you will appreciate later on.

Substituting $t = 1$ gives

$$G(1) = 0.323 + 0.387 + 0.197 + 0.080 + 0.006 + 0.006 = 1$$

as the sum of the probabilities has to be 1.

This can be set out in tabular form.

t^x	t^0	t^1	t^2	t^3	t^4	t^5
P(X = x)	0.322	0.387	0.197	0.080	0.007	0.007
t^xP(X = x)	0.323	$0.387 \times t$	$0.197 \times t^2$	$0.080 \times t^3$	$0.006 \times t^4$	$0.006 \times t^5$

Thus

$$G(t) = \sum_x t^x P(X = x)$$

and this can also be written as $E(t^x)$

which gives the general definition for a probability generating function.

The Poisson distribution

The way in which G(t) generates probabilities is best seen from a *theoretical* model. From the shape of the probability distribution for goals scored by away teams, it looks as though a Poisson distribution might produce a good fit.

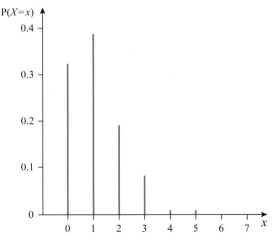

▲ **Figure 5.1**

Answers to exercises are available at www.hoddereducation.com/cambridgeextras

Using a suitable χ^2 test you can check that, at the 5% level of significance, a Poisson distribution with mean 1.08 (using the sample mean as an unbiased estimate of the population mean) produces a good fit. The *theoretical* probabilities for a Poisson distribution, with mean = 1.08, are given by

$$P(X = x) = e^{-1.08} \frac{1.08^x}{x!}.$$

Therefore the PGF G(t), given by $E(t^x) = \sum_x t^x P(X = x)$ may be written in a convenient form

$$G(t) = \sum_{x=0}^{\infty} e^{-1.08} \frac{1.08^x}{x!} t^x$$

When this is written as a series it is

$$G(t) = e^{-1.08}(1 + 1.08t + \frac{1.08^2}{2!}t^2 + \frac{1.08^3}{3!}t^3 + \ldots)$$

This can be written in the very neat and convenient algebraic form as

$$G(t) = e^{-1.08} e^{1.08t}$$

$$= e^{1.08(t-1)}$$

This result shows why the PGF can be useful; using a lot of probabilities as the coefficients of powers of t in a function does not at first seem like a good idea – and does not seem to do anything that listing the probabilities in a table could not do more easily. The reason that it can be worthwhile is that the function made up by using a list of probabilities as the coefficients of powers of t can sometimes be written in an alternative, simpler way. Here, by recognising an exponential series, the function, which started out as an infinite sum, can be written in a very neat way; this is the beautiful PGF trick at work.

Note

The PGF for a discrete random variable X, where $X \sim Po(\lambda)$, is given by $G(t) = e^{\lambda(t-1)}$. A formal proof is given later in this chapter.

Basic properties of probability generating functions

The example above illustrates some basic properties of probability generating functions (PGFs) for any discrete random variable X.

Definition $\qquad$ $G(t) = E(t^x) = \sum_x t^x P(X = x)$

$\Rightarrow$ $\qquad$ $G(1) = \sum_x 1^x P(X = x) = \sum_x P(X = x) = 1$

If X takes non-negative integral values only, then the probability generating function takes the form of a polynomial in t, which is often written as

$$G(t) = \sum_x p_x t^x = p_0 + p_1 t + p_2 t^2 + \ldots + p_n t^n + \ldots$$

where p_x denotes $P(X = x)$; for example, p_2 denotes $P(X = 2)$. The function G(t) may be either a finite or infinite polynomial: that is, it may terminate after n terms or continue indefinitely.

Probability generating functions are used for discrete random variables. The equivalent for continuous variables are *moment generating functions* but these are beyond the scope of this book.

The uniform distribution

Example 5.1

Let X be the discrete random variable that denotes the score when a fair die is thrown. Find the probability generating function (PGF) for X.

Solution

The probability distribution is given by

x	1	2	3	4	5	6
$P(X=x)$	$\frac{1}{6}$	$\frac{1}{6}$	$\frac{1}{6}$	$\frac{1}{6}$	$\frac{1}{6}$	$\frac{1}{6}$

The PGF for X is

$$G(t) = E(t^x) = \frac{1}{6}t + \frac{1}{6}t^2 + \frac{1}{6}t^3 + \frac{1}{6}t^4 + \frac{1}{6}t^5 + \frac{1}{6}t^6$$

$$= \frac{1}{6}t\left(1 + t + t^2 + t^3 + t^4 + t^5\right) \longleftarrow$$

The first six terms of a G.P. with first term 1 and common ratio t.

$$= \frac{1}{6}t \times \frac{1-t^6}{1-t}$$

$$= \frac{t\left(1-t^6\right)}{6\left(1-t\right)}$$

The binomial distribution

Example 5.2

Wei is a keen archer. He has taken part in many competitions. When aiming for the centre of the target, called the gold, he finds that he hits the gold 70% of the time. Find the PGF for X, the number of gold hits in four consecutive shots at the target.

Solution

As $X \sim B(4, 0.7)$, the probabilities are given by $P(X = x) = \binom{4}{x} 0.7^x (1 - 0.7)^{4-x}$, which gives the following distribution.

X	0	1	2	3	4
$P(X=x)$	0.3^4	$4 \times 0.7 \times 0.3^3$	$6 \times 0.7^2 \times 0.3^2$	$4 \times 0.7^3 \times 0.3$	0.7^4

The PGF for X is

$$G(t) = E(t^x) = \sum_{x=0}^{4} p_x t^x = \sum_{x=0}^{4} \binom{4}{x} \times 0.7^x \times 0.3^{4-x} \times t^x$$

$$= 0.3^4 + 4 \times 0.3^3 \times 0.7t + 6 \times 0.3^2 \times 0.7^2 t^2 + 4 \times 0.3 \times 0.7^3 t^3 + 0.7^4 t^4$$

$$= (0.3 + 0.7t)^4$$

Answers to exercises are available at www.hoddereducation.com/cambridgeextras

> ### Note
>
> When $X \sim B(n, p)$, i.e. $P(X = x) = \binom{n}{x} p^x q^{n-x}$, where $q = 1 - p$, the PGF for
>
> X is given by $G(t) = (q + pt)^n$. A formal proof is given later in this chapter.

The geometric distribution

Example 5.3

A pack of playing cards contains 52 cards, with 13 cards of each of four suits (hearts, spades, diamonds and clubs). A card is selected at random from a normal pack of cards. If it is a heart, the experiment ends, otherwise it is replaced, the pack is shuffled and another card is selected. Let X represent the number of cards selected up to and including the first heart. Find the PGF for X.

Solution

As X follows a geometric distribution, that is, $X \sim \text{Geo}\left(\frac{1}{4}\right)$, the probabilities are given by $P(X = x) = \frac{1}{4} \times \left(1 - \frac{1}{4}\right)^{x-1}$, which gives the following distribution.

X	1	2	3	4	5	...
$P(X = x)$	$\frac{1}{4}$	$\frac{1}{4} \times \frac{3}{4}$	$\frac{1}{4} \times \left(\frac{3}{4}\right)^2$	$\frac{1}{4} \times \left(\frac{3}{4}\right)^3$	$\frac{1}{4} \times \left(\frac{3}{4}\right)^4$	...

The PGF for X is

$$G(t) = E(t^x) = \sum_{x=1}^{\infty} p_x t^x = \sum_{x=1}^{\infty} \left(\frac{1}{4}\right)\left(\frac{3}{4}\right)^{x-1} t^x$$

$$= \left(\frac{1}{4}\right)t + \left(\frac{1}{4}\right)\left(\frac{3}{4}\right)t^2 + \left(\frac{1}{4}\right)\left(\frac{3}{4}\right)^2 t^3 + \left(\frac{1}{4}\right)\left(\frac{3}{4}\right)^3 t^4 + \left(\frac{1}{4}\right)\left(\frac{3}{4}\right)^4 t^5 + \ldots$$

$$= \left(\frac{1}{4}\right)t\left[1 + \left(\frac{3}{4}t\right) + \left(\frac{3}{4}t\right)^2 + \left(\frac{3}{4}t\right)^3 + \left(\frac{3}{4}t\right)^4 + \ldots\right]$$

$$= \left(\frac{1}{4}\right)t\left[\frac{1}{1 - \frac{3}{4}t}\right]$$

The sum to infinity of a G.P. with first term 1 and common ratio $\frac{3}{4}$.

$$= \frac{t}{4 - 3t}$$

> ### Note
>
> When $X \sim \text{Geo}(p)$, i.e. $P(X = x) = pq^{x-1}$, where $q = 1 - p$, the PGF for X is
>
> given by $G(t) = \dfrac{pt}{1 - qt}$. A formal proof is given later in this chapter.

Exercise 5A

1 Let X be the discrete random variable that denotes the sum of the scores when two fair dice are thrown. Construct a table for the probability distribution and so write down the PGF.

2 Let X be the discrete random variable that denotes the absolute difference of the scores when two fair dice are thrown ($x = 0, 1, 2, 3, 4, 5$). Construct a table for the probability distribution and so write down the PGF.

3 A fair coin is tossed three times and the number of tails appearing is noted as the discrete random variable X ($x = 0, 1, 2, 3$). Construct a table for the probability distribution and so write down the PGF.

4 A box contains three red balls and two green balls. They are taken out one at a time, *without* replacement. Let X represent the number of withdrawals until a red ball is chosen. Find its PGF.

5 The experiment in question **4** is repeated, but this time *with* replacement. Find the PGF for the number of withdrawals until a red ball is chosen.

6 A random number generator in a computer game produces values that can be modelled by the discrete random variable X with probability distribution given by

$P(X = r) = kr!$ $r = 1, 2, 3, 4, 5$.

Determine the value of k and so write down G(t), the PGF for X.

7 A student is shown three graphs. She is also given three equations, one for each graph. She matches each graph with its equation at random. Construct a table for the probability distribution of X, the number of correctly identified graphs. Deduce the PGF for X.

8 Two students are to be chosen to represent a class containing nine boys and six girls. Assuming that the students are chosen at random, find the PGF for X, the number of girls representing the group.

5.2 Expectation and variance

For any discrete probability distribution, the **expectation** (mean) and **variance** may be computed as follows, using the shorthand $P(X = x) = p_x$.

$$\mu = E(X) = \sum_x x p_x$$

$$\sigma^2 = \mathrm{Var}(X) = E(X - \mu)^2 = \sum_x (x - \mu)^2 p_x = \sum_x x^2 p_x - \mu^2$$

$$= E(X^2) - \mu^2.$$

By successively differentiating G(t), the probability generating function for X, formulae for the expectation and variance can be derived elegantly. It is here that the power of the PGF as an algebraic tool becomes apparent.

Answers to exercises are available at <u>www.hoddereducation.com/cambridgeextras</u>

Expectation

This may be obtained from the probability generating function $G(t)$ by differentiating with respect to t and evaluating the expression at $t = 1$.

$$G(t) = E(t^X) = \sum_x p_x t^x$$

$$\Rightarrow \qquad G'(t) = \sum_x x p_x t^{x-1} = E(Xt^{X-1})$$

$$\Rightarrow \qquad G'(1) = \sum_x x p_x = E(X)$$

Therefore $\mu = E(X) = G'(1)$.

Variance

This may be obtained by differentiating $G'(t)$ and evaluating the expression at $t = 1$.

$$G'(t) = \sum_x x p_x t^{x-1} = E(Xt^{X-1})$$

$$\Rightarrow \qquad G''(t) = \sum_x x(x-1) p_x t^{x-2} = E(X(X-1)t^{X-2})$$

$$\Rightarrow \qquad G''(1) = \sum_x x(x-1) p_x = E(X(X-1))$$

$$\Rightarrow \qquad G''(1) = \sum_x x^2 p_x - \sum_x x p_x = E(X^2) - E(X)$$

$$\Rightarrow \qquad E(X^2) = G''(1) + E(X) = G''(1) + G'(1)$$

Therefore

$$\sigma^2 = G''(1) + G'(1) - (G'(1))^2 \quad \text{or} \quad \sigma^2 = G''(1) + \mu - \mu^2.$$

Summary

Whenever a PGF is given for a discrete random variable, the expectation (mean) and variance may be evaluated using the following definitions.

$$\mu = E(X) = G'(1) \quad \textbf{and} \quad \sigma^2 = \text{Var}(X) = G''(1) + G'(1) - (G'(1))^2$$

In the following examples, the first one shows how the polynomial form for a probability generating function $G(t)$ may be differentiated twice in order to derive the expectation and variance. You should notice that this just reproduces exactly the usual calculations for mean and variance, and there is no gain from the PGF method. However, the two further examples demonstrate the remarkable power of the PGF method when it is possible to write the function $G(t)$ in a simpler form.

Example 5.4

Let X be the discrete random variable that denotes the absolute difference of scores when two fair dice are thrown. Using a PGF, confirm that $E(X)$ is just under 2 and determine $\text{Var}(X)$.

Solution

The probability distribution of X was derived in Exercise 5A, question 2, as

x	0	1	2	3	4	5
$P(X = x)$	$\dfrac{3}{18}$	$\dfrac{5}{18}$	$\dfrac{4}{18}$	$\dfrac{3}{18}$	$\dfrac{2}{18}$	$\dfrac{1}{18}$

The PGF for X is as follows.

$$G(t) = \frac{3}{18} + \frac{5}{18}t + \frac{4}{18}t^2 + \frac{3}{18}t^3 + \frac{2}{18}t^4 + \frac{1}{18}t^5$$

$$\Rightarrow \quad G'(t) = \frac{5}{18} + \frac{8}{18}t + \frac{9}{18}t^2 + \frac{8}{18}t^3 + \frac{5}{18}t^4$$

$$\Rightarrow \quad G''(t) = \frac{8}{18} + \frac{18}{18}t + \frac{24}{18}t^2 + \frac{20}{18}t^3$$

Therefore $\quad G'(1) = \frac{5}{18} + \frac{8}{18} + \frac{9}{18} + \frac{8}{18} + \frac{5}{18} = \frac{35}{18} = 1\frac{17}{18}$

and $\quad G''(1) = \frac{8}{18} + \frac{18}{18} + \frac{24}{18} + \frac{20}{18} = \frac{70}{18} = 3\frac{8}{9}.$

$$\Rightarrow \quad E(X) = G'(1) = 1\frac{17}{18}$$

and $\quad \text{Var}(X) = G''(1) + G'(1) - (G'(1))^2$

$$= 3\frac{8}{9} + 1\frac{17}{18} - \left(1\frac{17}{18}\right)^2 = 2\frac{17}{324} \approx 2.05$$

> ❯ Confirm that these results may also be derived from first principles, i.e.
> $\mu = E(X) = \sum_x x p_x$ and $\sigma^2 = \text{Var}(X) = \sum_x x^2 p_x - \mu^2.$

Example 5.5

Four unbiased coins are tossed and the number of heads (X) is noted. Show that the PGF for X is given by $G(t) = \frac{1}{16}(1+t)^4$. From this, calculate the mean and variance of X.

Solution

As $X \sim B(4, \frac{1}{2})$, the probabilities are given by

$$P(X = x) = \binom{4}{x}(\tfrac{1}{2})^x (\tfrac{1}{2})^{4-x} = \binom{4}{x}(\tfrac{1}{2})^4 = \binom{4}{x}\tfrac{1}{16}$$

The PGF for X is

$$G(t) = E(t^x) = \sum_{x=0}^{4} p_x t^x = \tfrac{1}{16}\sum_{x=0}^{4}\binom{4}{x}t^x$$

$$= \tfrac{1}{16}(1 + 4t + 6t^2 + 4t^3 + t^4)$$

$$= \tfrac{1}{16}(1 + t)^4$$

Answers to exercises are available at www.hoddereducation.com/cambridgeextras

From the PGF the mean and variance of X are found by differentiation.

$$G'(t) = 4 \times \frac{1}{16}(1+t)^3 = \frac{1}{4}(1+t)^3 \quad \Rightarrow \quad G'(1) = \frac{1}{4} \times 2^3 = 2$$

$$G''(t) = 3 \times \frac{1}{4}(1+t)^2 = \frac{3}{4}(1+t)^2 \quad \Rightarrow \quad G''(1) = \frac{3}{4} \times 2^2 = 3$$

$$\Rightarrow \qquad E(X) = G'(1) = 2$$

and $\qquad \mathrm{Var}(X) = G''(1) + G'(1) - (G'(1))^2 = 3 + 2 - 2^2 = 1$

Example 5.6

A box contains three red balls and two green balls. They are taken out one at a time, with replacement. Let X represent the number of withdrawals until a red ball is chosen. Calculate the mean and variance of X.

Solution

In this experiment, the outcomes follow a geometric distribution.

$X \sim \mathrm{Geo}(p)$, where $p = \frac{3}{5}$ and $q = 1 - \frac{3}{5} = \frac{2}{5}$, and so the PGF is

$$G(t) = \frac{pt}{1-qt} = \frac{\frac{3}{5}t}{1 - \frac{2}{5}t} = \frac{3t}{5 - 2t}.$$

Differentiating with respect to t, using the quotient rule:

$$G'(t) = \frac{(5-2t) \times 3 - 3t \times (-2)}{(5-2t)^2} = \frac{15}{(5-2t)^2} \quad \Rightarrow \quad G'(1) = \frac{15}{9} = 1\frac{2}{3}$$

Differentiating with respect to t, using the chain rule:

$$G''(t) = -2 \times 15 \times (5-2t)^{-3} \times (-2) = \frac{60}{(5-2t)^3} \quad \Rightarrow \quad G''(1) = \frac{60}{27} = 2\frac{2}{9}$$

$$\Rightarrow \qquad E(X) = G'(1) = 1\frac{2}{3}$$

and $\qquad \mathrm{Var}(X) = G''(1) + G'(1) - \left(G'(1)\right)^2 = 2\frac{2}{9} + 1\frac{2}{3} - \left(1\frac{2}{3}\right)^2 = 1\frac{1}{9}$

Example 5.7	A discrete random variable X ($x = 0, 1, 2$) has PGF given by $G(t) = a + bt + ct^2$, where a, b and c are constants. If the mean is $1\frac{1}{4}$ and the variance is $\frac{11}{16}$, find the values of a, b and c.

Solution

As $P(X = 0) = a$, $P(X = 1) = b$ and $P(X = 2) = c$

$$G(t) = a + bt + ct^2 \qquad \Rightarrow \qquad G(1) = a + b + c = 1 \qquad ①$$

As $\mu = 1\frac{1}{4}$ and $E(X) = G'(1)$:

$$G'(t) = b + 2ct \qquad \Rightarrow \qquad G'(1) = b + 2c = 1\frac{1}{4} \qquad ②$$

As $\sigma^2 = G''(1) + G'(1) - (G'(1))^2$, so $G''(1) = \sigma^2 - G'(1) + (G'(1))^2$

$$= \frac{11}{16} - 1\frac{1}{4} + \left(1\frac{1}{4}\right)^2 = 1$$

$$G''(t) = 2c \qquad \Rightarrow \qquad G''(1) = 2c = 1 \qquad ③$$

From equation ③: $\qquad 2c = 1 \qquad \Rightarrow \qquad c = \frac{1}{2}$

From equation ②: $\qquad b + 2c = 1\frac{1}{4} \qquad \Rightarrow \qquad b = 1\frac{1}{4} - 2 \times \frac{1}{2} = \frac{1}{4}$

From equation ①: $\qquad a + b + c = 1 \qquad \Rightarrow \qquad a = 1 - \frac{1}{4} - \frac{1}{2} = \frac{1}{4}$

Therefore $\qquad a = \frac{1}{4}, b = \frac{1}{4}, c = \frac{1}{2}.$

Exercise 5B

1. A random variable has PGF $G(t) = \frac{1}{6} + \frac{1}{3}t + \frac{1}{4}t^3 + \frac{1}{6}t^4 + \frac{1}{12}t^5$. Find its mean and variance.

2. A random variable X has PGF $G(t) = \frac{1}{81}(1 + 2t)^4$. Calculate $E(X)$ and $Var(X)$.

3. Let X be the discrete random variable that denotes the score when a fair die is thrown. Use its PGF to find $E(X)$ and $Var(X)$.

4. A fair coin is tossed three times and the number of tails appearing is noted as the discrete random variable $X(x = 0, 1, 2, 3)$. Use the PGF

$$G(t) = \frac{1}{8}\left(1 + 3t + 3t^2 + t^3\right)$$

   to calculate the mean and variance of X.

5. The probability generating function of a discrete random variable, X, is $G(t) = (kt + 1 - k)^n$, where k is a constant ($0 \leqslant k \leqslant 1$) and n is a positive integer.

   Prove that $E(X) = nk$, and find the variance of X.

M 6. An ordinary pack of 52 playing cards containing four ace cards is cut four times. Let X represent the number of aces appearing. Find the probability generating function for X and so find the mean and variance of X.

M 7. A king and queen want a son and heir to their kingdom. Let X represent the number of children they have until a boy is born.

Answers to exercises are available at www.hoddereducation.com/cambridgeextras

(i) Assuming that each pregnancy results in a single child, and that the probability of a boy is 0.5, show that the PGF for X is given by

$$G(t) = \frac{t}{2-t}$$

(ii) From this show that both $E(X)$ and $\text{Var}(X) = 2$.

CP 8 A random variable has PGF $G(t) = a + bt + ct^2$. It has mean $\frac{7}{6}$ and variance $\frac{29}{36}$. Find the values for a, b and c.

9 Two fair dice are thrown and X represents the larger of the two scores. Find the PGF for X and so find the mean and variance of X.

Landmark competition

In a competition, the entrant has to match a number of famous landmarks (A, B, C, etc.) with the country in which it is to be found (1, 2, 3, etc.). The number of countries to choose from is the same as the number of landmarks.

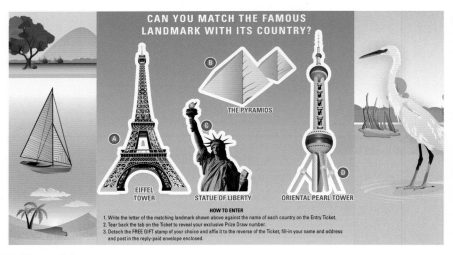

▲ Figure 5.2

Let X represent the number of correct matches that the entrant makes. For the purposes of this investigation, assume that the matching of the landmark to its country is chosen at random.

In each case take the correct matching as A ↔ 1, B ↔ 2, C ↔ 3, etc.

1 For three famous landmarks the possible matchings and values of X are as follows.

A	1	1	2	2	3	3
B	2	3	1	3	1	2
C	3	2	3	1	2	1
X	3	1	1	0	0	1

Find the PGF for X and show the $E(X) = \text{Var}(X) = 1$.

2   (i)   For *four* famous landmarks explain why there are 4! = 24 possible matchings, only one of which is a complete matching.

     (ii) Show that the PGF for X is given by $G(t) = \frac{3}{8} + \frac{1}{3}t + \frac{1}{4}t^2 + \frac{1}{24}t^4$.

     (iii) Use the PGF to show that $E(X) = Var(X) = 1$.

3   (i)   For n $(n > 2)$ famous landmarks there are $n!$ possible matchings. Explain why $P(X = n) = \frac{1}{n!}$ and $P(X = n - 1) = 0$.

     (ii) Prove that $P(X = n - 2) = \frac{1}{2 \times (n - 2)!}$ and $P(X = n - 3) = \frac{1}{3 \times (n - 3)!}$

     (iii) Verify these results for $n = 3$ and $n = 4$.

4   For *five* famous landmarks, use the results from part **3** and the assumption that $E(X) = 1$ to derive the PGF for X.

5   For *six* famous landmarks, use the results from part **3** and the assumptions that $E(X) = 1$ and $Var(X) = 1$ to derive the PGF for X.

6   (i)   Compare the PGFs in parts **1, 2, 4** and **5**. How are they related to each other?

     (ii) How can you use the PGF for n possible matches to derive the PGF for $n - 1$ possible matches and for $n + 1$ possible matches?

5.3 The sum of independent random variables

At the beginning of the chapter you saw that, with a mean of 1.08 goals per match, the random variable X, the number of goals scored by the *away* team in each game was distributed approximately Po(1.08). So the probability generating function for X is given by $G_X(t) = e^{1.08(t - 1)}$.

Let Y represent the number of goals scored by the *home* team. Then, from the data collected for the same season for *home* teams:

Number of goals	Frequency	Relative frequency
0	100	0.216…
1	157	0.339…
2	110	0.238…
3	53	0.114…
4	28	0.060…
5	10	0.021…
6	3	0.006…
7	1	0.002…

the sample mean is 1.56 (to 3 s.f.). Again a Poisson distribution is a good fit to model the number of *home* goals: that is, $Y \sim$ Po(1.56) with PGF $G_Y(t) = e^{1.56(t - 1)}$.

Answers to exercises are available at www.hoddereducation.com/cambridgeextras

What do the models predict for the distribution of the total number of goals scored by both sides in a match? For instance, what do they predict for $P(X + Y = 3)$?

$$P(X + Y = 3) = P(X = 0 \text{ and } Y = 3) + P(X = 1 \text{ and } Y = 2) +$$
$$P(X = 2 \text{ and } Y = 1) + P(X = 3 \text{ and } Y = 0).$$

The probabilities of $P(X = a \text{ and } Y = b)$ are not simply related to the separate probabilities of away and home goals, $P(X = a)$ and $P(Y = b)$, unless you assume that the random variables X and Y are independent. You might not expect this to be the case (if one side plays a lot better and scores a lot of goals, then their opponents do not) but, surprisingly, the figures suggest that it is true to a reasonable approximation.

If X and Y are independent, $P(X = a \text{ and } Y = b) = P(X = a) \times P(Y = b)$, so

$$P(X + Y = 3) = P(X = 0) \times P(Y = 3) + P(X = 1) \times P(Y = 2) +$$
$$P(X = 2) \times P(Y = 1) + P(X = 3) \times P(Y = 0).$$

This figure can be found very neatly from the PGFs for X and Y, because

$$G_X(t) \times G_Y(t) = (P(X = 0) + P(X = 1)t + P(X = 2)t^2 +$$
$$P(X = 3)t^3 + \ldots) \times (P(Y = 0)$$
$$+ P(Y = 1)t + P(Y = 2)t^2 + P(Y = 3)t^3 + \ldots).$$

Picking from this product the terms in t^3, you get

$$P(X = 0) \times P(Y = 3)t^3 + P(X = 1)t \times P(Y = 2)t^2 + P(X = 2)t^2 \times$$
$$P(Y = 1)t + P(X = 3)t^3 \times P(Y = 0)$$
$$= P(X + Y = 3) \times t^3.$$

The same relationship holds for other values of $X + Y$, and so for other powers of t in the product $G_X(t) \times G_Y(t)$. This means that

$$G_X(t) \times G_Y(t) = P(X + Y = 0) + P(X + Y = 1)t +$$
$$P(X + Y = 2)t^2 + P(X + Y = 3)t^3 + \ldots,$$

but the right-hand side of this equation is just the PGF for the random variable $X + Y$, so that

$$G_X(t) \times G_Y(t) = G_{X+Y}(t)$$

and so

$$G_{X+Y}(t) = e^{1.08(t-1)} \times e^{1.56(t-1)} = e^{2.64(t-1)}.$$

This PGF is just that for a Poisson random variable with mean 2.64, which is the sum of the means for the two independent random variables that you are adding. The result, that the sum of two independent Poisson variables with means λ and μ is a Poisson variable with mean $\lambda + \mu$, should be familiar to you from your earlier work on the Poisson distribution.

The result derived from these variables in fact holds, by exactly the same argument, for any pair of independent random variables X and Y. A formal proof is given here.

The coefficient of t^r in $(a_0 + a_1 t + a_2 t^2 + \ldots + a_r t^r + \ldots)(b_0 + b_1 t + b_2 t^2 + \ldots + b_r t^r + \ldots)$ is $a_0 b_r + a_1 b_{r-1} + \ldots + a_s b_{r-s} + \ldots a_r b_0$. If the two power series represent $G_X(t)$ and $G_Y(t)$, respectively, then $a_s = P(X = s)$ and $b_s = P(Y = s)$ for each s, so that the coefficient of t^r in $G_X(t) \times G_Y(t)$ can be written as

$$\sum_{s=0}^{r} a_s b_{r-s} = \sum_{s=0}^{r} P(X = s) \times P(Y = r - s)$$

$$= \sum_{s=0}^{r} P(X = s \text{ and } Y = r - s)$$

$$= P(X + Y = r). \longleftarrow \boxed{\text{since } X \text{ and } Y \text{ are independent}}$$

Therefore

$$G_X(t) \times G_Y(t) = \sum_{r=0}^{\infty} P(X + Y = r) t^r = G_{X+Y}(t).$$

A neater argument is as follows.

$$G_{X+Y}(t) \equiv E(t^{X+Y}) = E(t^X t^Y) = E(t^X) \times E(t^Y)$$

$$= G_X(t) \times G_Y(t) \longleftarrow \boxed{\text{since } X \text{ and } Y \text{ are independent}}$$

However, this relies on the result that if X and Y are independent, then

$$E(f(X) \times g(Y)) = E(f(X)) \times E(g(Y))$$

for any functions f and g, which is not proved in this course.

> If X and Y are two *independent* discrete random variables with PGFs $G_X(t)$ and $G_Y(t)$, then the probability generating function for $X + Y$ is given by
>
> $G_{X+Y}(t) = G_X(t) \times G_Y(t)$
>
> i.e. the PGF of the sum = the product of the PGFs.
>
> This is known as the **convolution theorem**.

Example 5.8

A discrete random variable X ($x = 0, 1, 2$) has PGF $G_X(t) = \frac{1}{4} + \frac{1}{4}t + \frac{1}{2}t^2$ and another discrete random variable Y ($y = 0, 1$) has PGF $G_Y(t) = \frac{2}{3} + \frac{1}{3}t$.

Assume that X and Y are independent.

(i) Find the PGF for $X + Y$, i.e. $G_{X+Y}(t)$.

(ii) Show that $E(X + Y) = E(X) + E(Y)$ and $Var(X + Y) = Var(X) + Var(Y)$.

Solution

(i) $G_{X+Y}(t) = G_X(t) \times G_Y(t) = \left(\frac{1}{4} + \frac{1}{4}t + \frac{1}{2}t^2\right)\left(\frac{2}{3} + \frac{1}{3}t\right)$

$$= \frac{1}{4} \times \frac{2}{3} + \left(\frac{1}{4} \times \frac{1}{3} + \frac{1}{4} \times \frac{2}{3}\right)t + \left(\frac{1}{4} \times \frac{1}{3} + \frac{1}{2} \times \frac{2}{3}\right)t^2 + \left(\frac{1}{2} \times \frac{1}{3}\right)t^3$$

$$= \frac{1}{6} + \frac{1}{4}t + \frac{5}{12}t^2 + \frac{1}{6}t^3$$

5

Answers to exercises are available at www.hoddereducation.com/cambridgeextras

(ii) By differentiation

$$G'_X(t) = \frac{1}{4} + t, G'_Y(t) = \frac{1}{3} \quad \text{and} \quad G'_{X+Y}(t) = \frac{1}{4} + \frac{5}{6}t + \frac{1}{2}t^2$$

$$\Rightarrow \quad E(X) = G'_X(1) = \frac{1}{4} + 1 = 1\frac{1}{4} \quad \text{and} \quad E(Y) = G'_Y(1) = \frac{1}{3}$$

$$E(X+Y) = G'_{X+Y}(1) = \frac{1}{4} + \frac{5}{6} + \frac{1}{2} = 1\frac{7}{12}$$

$$\Rightarrow \quad E(X) + E(Y) = 1\frac{1}{4} + \frac{1}{3} = 1\frac{7}{12} = E(X+Y)$$

By further differentiation

$$G''_X(t) = 1, G''_Y(t) = 0 \quad \text{and} \quad G''_{X+Y}(t) = \frac{5}{6} + t$$

$$\text{Var}(X) = G''_X(1) + G'_X(1) - (G'_X(1))^2 = 1 + 1\frac{1}{4} - \left(1\frac{1}{4}\right)^2 = \frac{11}{16}$$

$$\text{Var}(Y) = G''_Y(1) + G'_Y(1) - (G'_Y(1))^2 = 0 + \frac{1}{3} - \left(\frac{1}{3}\right)^2 = \frac{2}{9}$$

$$\text{Var}(X+Y) = G''_{X+Y}(1) + G'_{X+Y}(1) - \left(G'_{X+Y}(1)\right)^2$$

$$= 1\frac{5}{6} + 1\frac{7}{12} - \left(1\frac{7}{12}\right)^2 = \frac{131}{144}$$

$$\Rightarrow \quad \text{Var}(X) + \text{Var}(Y) = \frac{11}{16} + \frac{2}{9} = \frac{131}{144} = \text{Var}(X+Y)$$

Extension to three or more random variables

The result, that the PGF of the sum of two independent random variables is the product of the PGFs, can be extended to three or more variables. For example, if X, Y and Z are three independent discrete random variables with PGFs $G_X(t)$, $G_Y(t)$ and $G_Z(t)$, then the probability generating function for $X + Y + Z$ is given by

$$G_{X+Y+Z}(t) = G_X(t) \times G_Y(t) \times G_Z(t).$$

If n independent discrete random variables all have the same PGF, $G(t)$, then the probability generating function for their sum is

$$(G(t))^n.$$

Example 5.9

Find the probability generating function for the total number of sixes when five fair dice are thrown. Deduce the mean and variance.

Solution

When *one* die is thrown, a 6 occurs with probability $\frac{1}{6}$, therefore the PGF for the number of sixes is $G(t) = \frac{5}{6} + \frac{1}{6}t$. Therefore, when *five* dice are thrown, the PGF for X, the total number of sixes is given by

$$G_X(t) = (G(t))^5 = \left(\frac{5}{6} + \frac{1}{6}t\right)^5$$

Applying the results derived earlier.

$$G'_X(t) = 5 \times \left(\frac{5}{6} + \frac{1}{6}t\right)^4 \times \frac{1}{6} = \frac{5}{6}\left(\frac{5}{6} + \frac{1}{6}t\right)^4$$

$$G''_X(t) = 4 \times \frac{5}{6}\left(\frac{5}{6} + \frac{1}{6}t\right)^3 \times \frac{1}{6} = \frac{5}{9}\left(\frac{5}{6} + \frac{1}{6}t\right)^3$$

$$\Rightarrow \quad E(X) = G'(1) = \frac{5}{6}\left(\frac{5}{6} + \frac{1}{6}\right)^4 = \frac{5}{6} \qquad \left(\text{as } \frac{5}{6} + \frac{1}{6} = 1\right)$$

and

$$\text{Var}(X) = G''(1) + G'(1) - (G'(1))^2$$

$$= \frac{5}{9}\left(\frac{5}{6} + \frac{1}{6}\right)^3 + \frac{5}{6} - \left(\frac{5}{6}\right)^2 = \frac{5}{9} + \frac{5}{6} - \frac{25}{36} = \frac{25}{36}$$

Note

As $X \sim B(5, \frac{1}{6})$, the results that $E(X) = 5 \times \frac{1}{6}$ and $\text{Var}(X) = 5 \times \frac{1}{6} \times \frac{5}{6}$ have been verified. The proof for the mean and variance of $X \sim B(n, p)$ is given later in this chapter.

Example 5.10

Gina regularly practises archery. The probability of her hitting the gold (the centre of the target) with any arrow is $\frac{1}{3}$.

(i) Find the PGF $G(t)$ of the number of shots until she hits the gold for the first time.

(ii) Show that the PGF of the number of shots until she hits the gold for the second time is $(G(t))^2$.

(iii) Explain why the PGF of the number of shots until she hits the gold for the kth time is $(G(t))^k$.

Solution

(i) Let X represent the number of shots up to and including Gina's *first* gold, then the random variable X forms a geometric distribution with

$$P(X = r) = \frac{1}{3}\left(\frac{2}{3}\right)^{r-1}$$

The PGF for X is

$$G_X(t) = \left(\frac{1}{3}\right)t + \left(\frac{1}{3}\right)\left(\frac{2}{3}\right)t^2 + \left(\frac{1}{3}\right)\left(\frac{2}{3}\right)^2 t^3 + \left(\frac{1}{3}\right)\left(\frac{2}{3}\right)^3 t^4 + \left(\frac{1}{3}\right)\left(\frac{2}{3}\right)^4 t^5 + \dots$$

$$= \left(\frac{1}{3}\right)t\left[1 + \left(\frac{2}{3}t\right) + \left(\frac{2}{3}t\right)^2 + \left(\frac{2}{3}t\right)^3 + \left(\frac{2}{3}t\right)^4 + \dots\right]$$

$$= \left(\frac{1}{3}\right)t\left[\frac{1}{1 - \frac{2}{3}t}\right]$$

$$= \frac{t}{3 - 2t}$$

Answers to exercises are available at www.hoddereducation.com/cambridgeextras

(ii) Let Y represent the number of shots up to and including Gina's *second* gold. This is called a **negative binomial distribution**, given by the formula

$$P(Y = r) = \begin{pmatrix} r-1 \\ 1 \end{pmatrix} \left(\frac{1}{3}\right)^2 \left(\frac{2}{3}\right)^{r-2} = (r-1)\left(\frac{1}{3}\right)^2 \left(\frac{2}{3}\right)^{r-2}$$

The PGF for Y is

$$G_Y(t) = \left(\frac{1}{3}\right)^2 t^2 + 2\left(\frac{1}{3}\right)^2 \left(\frac{2}{3}\right)t^3 + 3\left(\frac{1}{3}\right)^2 \left(\frac{2}{3}\right)^2 t^4 + 4\left(\frac{1}{3}\right)^2 \left(\frac{2}{3}\right)^3 t^5 + \dots$$

$$= \left(\frac{1}{3}t\right)^2 \left[1 + 2\left(\frac{2}{3}t\right) + 3\left(\frac{2}{3}t\right)^2 + 4\left(\frac{2}{3}t\right)^3 + \dots\right]$$

$$= \left(\frac{1}{3}\right)t^2 \left[\frac{1}{(1 - \frac{2}{3}t)^2}\right]$$

$$= \left[\frac{t}{3 - 2t}\right]^2 = (G_X(t))^2$$

(iii) The number of shots required until the kth success is equivalent to the sum of k geometric distributions, so the PGF for this sum is the product of the PGFs, i.e. $(G(t))^k$.

5.4 The PGFs for some standard discrete probability distributions

There are several standard discrete probability distributions that are useful for modelling statistical data. Three of these, the binomial, the Poisson and the geometric distributions, you have met and used before. In this final section you will see how the PGF is derived for each of these discrete distributions and how it is used to provide neat proofs for the expectation, $E(X)$, and variance, $Var(X)$.

The binomial distribution

Let Y be the discrete random variable with probability distribution

y	0	1
$P(Y = y)$	q	p

The PGF for Y is therefore $G(t) = q + pt$ and $G(1) = q + p = 1$.

Now let the discrete random variable X be defined by $X = Y_1 + Y_2 + \dots + Y_n$, where each Y_i has PGF $G(t) = q + pt$. Then X has PGF $G_X(t)$, given by

$$G_X(t) = (G(t))^n = (q + pt)^n.$$

If '1' represents 'success' and '0' represents 'failure', then X represents the number of 'successes' in n independent trials, for which the probability of 'success' is p and the probability of failure is q.

Therefore $X \sim B(n, p)$ with PGF $G_X(t) = (q + pt)^n$.

Having established the PGF, the mean and variance may be proved.

Differentiating with respect to t

$$G'_X(t) = n(q + pt)^{n-1} \times p = np(q + pt)^{n-1}$$

$\Rightarrow$ $\qquad G'_X(1) = np(q + p)^{n-1} = np$

and $\qquad G''_X(t) = (n-1)np(q + pt)^{n-2} \times p = n(n-1)p^2(q + pt)^{n-2}$

$\Rightarrow$ $\qquad G''_X(1) = n(n-1)p^2(q + p)^{n-2} = n(n-1)p^2$

Therefore

$$E(X) = G'_X(1) = np$$

$$\text{Var}(X) = G''_X(1) + G'_X(1) - (G'_X(1))^2 = n(n-1)p^2 + np - (np)^2$$

$$= n^2p^2 - np^2 + np - n^2p^2$$

$$= np(1 - p) = npq$$

i.e. $\qquad E(X) = np$ and $\text{Var}(X) = npq$

The Poisson distribution

Let X be a Poisson random variable with parameter λ; that is, $X \sim \text{Po}(\lambda)$, then the PGF for X is

$$G_X(t) = e^{-\lambda} + e^{-\lambda} \times \lambda t + e^{-\lambda} \times \frac{\lambda^2}{2!}t^2 + \ldots + e^{-\lambda} \times \frac{\lambda^r}{r!}t^r + \ldots$$

$$= e^{-\lambda}\left(1 + \lambda t + \frac{(\lambda t)^2}{2!} + \ldots + \frac{(\lambda t)^r}{r!} + \ldots\right) = e^{-\lambda} \times e^{\lambda t} = e^{\lambda(t-1)}$$

Having established the PGF, the mean and variance may be proved.

Differentiating with respect to t

$$G'_X(t) = \lambda e^{\lambda(t-1)} \quad \Rightarrow \quad G'_X(1) = \lambda e^0 = \lambda$$

$$G''_X(t) = \lambda^2 e^{\lambda(t-1)} \quad \Rightarrow \quad G''_X(1) = \lambda^2 e^0 = \lambda^2$$

Therefore

$$E(X) = G'_X(1) = \lambda$$

$$\text{Var}(X) = G''_X(1) + G'_X(1) - (G'_X(1))^2 = \lambda^2 + \lambda - (\lambda)^2 = \lambda$$

i.e. $\qquad E(X) = \text{Var}(X) = \lambda$

Answers to exercises are available at www.hoddereducation.com/cambridgeextras

While this is a useful extension to the topic, it is beyond the requirements of the Cambridge International syllabus.

The Poisson approximation to the binomial

You may know that if $X \sim B(n, p)$, where n is large and p is small (for example, $n = 150$, $p = 0.02$), then a good approximation to the distribution of X is $X \sim Po(\lambda)$, where $\lambda = np$. The PGFs can be used to show that the Poisson distribution is the limit of the binomial distribution as $n \Rightarrow \infty$, $p \Rightarrow 0$ and np remains fixed.

For the binomial distribution $G(t) = (q + pt)^n$, where $q = 1 - p$

$$\Rightarrow \qquad G(t) = (1 - p + pt)^n = (1 + p(t - 1))^n.$$

But $\lambda = np \Rightarrow p = \lambda/n$, where λ is fixed

$$\Rightarrow \qquad G(t) = \left(1 + \frac{\lambda(t - 1)}{n}\right)^n.$$

Using the result that $e^x = \lim \left(1 + \frac{x}{n}\right)^n$ as $n \to \infty$, the limit of $G(t)$ as $n \to \infty$ is given by

$$G(t) = \lim \left(1 + \frac{\lambda(t - 1)}{n}\right)^n \text{ as } n \to \infty = e^{\lambda(t - 1)}.$$

The geometric distribution

If a sequence of independent trials is conducted, for each of which the probability of success is p and that of failure q (where $q = 1 - p$), and the random variable X is the number of the trial on which the first success occurs, then X has a geometric distribution with $P(X = r) = pq^{r - 1}, r \geq 1$.

The corresponding PGF is given by

$$G(t) = pt + pqt^2 + pq^2t^3 + \ldots + pq^{r - 1}t^r + \ldots$$

$$= pt[1 + qt + (qt)^2 + \ldots + (qt)^{r - 1} + \ldots]$$

$$= pt\left(\frac{1}{1 - qt}\right) = \frac{pt}{1 - qt}.$$

Differentiating with respect to t, using the quotient rule

$$G'(t) = \frac{(1 - qt) \times p - pt \times (-q)}{(1 - qt)^2} = \frac{p}{(1 - qt)^2} \quad \Rightarrow \quad G'(1) = \frac{p}{(1 - q)^2} = \frac{1}{p}.$$

Differentiating $G(t)$ with respect to t, using the chain rule

$$G''(t) = 2p(1 - qt)^{-3} \times (-q) = \frac{2pq}{(1 - qt)^3} \quad \Rightarrow \quad G''(1) = \frac{2pq}{(1 - q)^3} = \frac{2q}{p^2}$$

$$\Rightarrow \qquad E(X) = G'(1) = \frac{1}{p}$$

$$Var(X) = G''(1) + G'(1) - (G'(1))^2$$

$$= \frac{2q}{p^2} + \frac{1}{p} - \left(\frac{1}{p}\right)^2 = \frac{2q + p - 1}{p^2} = \frac{q}{p^2}$$

i.e $\qquad E(X) = \frac{1}{p}$ and $Var(X) = \frac{q}{p^2}$

Exercise 5C

1 (i) A fair die is thrown repeatedly until a 6 occurs. Show that the PGF for the number of throws required is given by

$$G(t) = \frac{t}{6 - 5t}$$

(ii) Obtain the PGF for the number of throws required to obtain two 6s (not necessarily consecutively).

(iii) Calculate the expected value and variance of the number of throws required to obtain two sixes.

2 The probability distributions of two independent random variables, X and Y, are as follows.

x	1	2	3
$P(X = x)$	0.6	0.4	
y	1	2	3
$P(Y = y)$	0.2	0.5	0.3

(i) Write down the PGFs for X and Y. Hence show that the PGF for $X + Y$ is given by $G(t) = t^2 (0.12 + 0.38t + 0.38t^2 + 0.12t^3)$.

(ii) Show that $E(X + Y) = E(X) + E(Y)$
and $Var(X + Y) = Var(X) + Var(Y)$.

3 In a traffic census on a two-way stretch of road, the number of vehicles per minute, X, travelling past a checkpoint in one direction is modelled by a Poisson distribution with parameter 3.4, and the number of vehicles per minute, Y, travelling in the other direction is modelled by a Poisson distribution with parameter 4.8, i.e. $X \sim Po(3.4)$ and $Y \sim Po(4.8)$.

(i) Find the PGFs for both X and Y.

(ii) Show that $E(X) = Var(X) = 3.4$.

(iii) Find the PGF for $X + Y$, the total number of vehicles passing the checkpoint per minute, assuming that X and Y are independent.

(iv) Demonstrate in two ways that $E(X + Y) = Var(X + Y) = 8.2$.

Answers to exercises are available at www.hoddereducation.com/cambridgeextras

PS 4 A game consists of rolling two dice, one six-sided and the other four-sided, and adding the scores together. Both dice are fair; the first is numbered 1 to 6 and the second is numbered 1 to 4.

Show that the PGF of Z, where Z is the sum of the scores on the two dice, is

$$G(t) = \frac{t^2}{24} \times \frac{(1-t^6)(1-t^4)}{(1-t)^2}.$$

5 (i) The random variable X can take values $1, 2, 3, \ldots$ and its PGF is $G(t)$. Show that the probability that X is even is given by $\frac{1}{2}(1 + G(-1))$.

(ii) For each of the following probability distributions, state the PGF and, from this, find the probability that the outcome is even.

(a) The total score when two ordinary dice are thrown.

(b) The number of attempts it takes Kerry to pass her driving test, given that the probability of passing at any attempt is $\frac{2}{3}$.

M 6 Two people, A and B, fire alternately at a target, the winner of the game being the first to hit the target. The probability that A hits the target with any particular shot is $\frac{1}{4}$ and the probability that B hits the target with any particular shot is $\frac{1}{5}$.

Given that A fires first, find:

(i) the probability that B wins the game with his first shot

(ii) the probability that A wins the game with his second shot

(iii) the probability that A wins the game.

(iv) The total number of shots fired by A and B is denoted by the random variable R. Show that the probability generating function of R is given by

$$G(t) = \frac{5t + 3t^2}{4(5 - 3t^{2)}}.$$

Find $E(R)$.

7 (i) The variable X has a Poisson distribution with mean λ. Write down the value of $P(X = r)$ and the PGF for X in the form $G_X(t) = \ldots$.

(ii) The variable Y can only take values $2, 3, 4, \ldots$ and is such that $P(Y = r) = kP(X = r)$, where k is a constant and $r > 1$. Find the value of k.

(iii) Show that the PGF of Y is given by

$$G(t) = \frac{e^{\lambda t} - \lambda t - 1}{e^\lambda - \lambda - 1}.$$

(iv) Calculate $E(Y)$ and show that $\text{Var}(Y) = \left(\frac{\mu}{\gamma}\right)(\lambda^2 e^\lambda + \gamma - \mu\gamma)$, where $\mu = E(Y)$ and $\gamma = \lambda(e^\lambda - 1)$.

M 8 Two duellists take alternate shots at each other until one of them scores a 'hit' on the other. If the probability that the first duellist scores a 'hit' is $\frac{1}{5}$ and the probability that the second scores a 'hit' is $\frac{1}{3}$, find the expected number of shots *before* a hit.

9 Independent trials, on each of which the probability of a 'success' is p $(0 < p < 1)$, are being carried out. The random variable, X, counts the number of trials up to *but not* including that on which the first 'success' is obtained.

(i) Write down an expression for $P(X = x)$ for $x = 0, 1, 2, \ldots$ and show that the probability generating function of X is

$$G(t) = \frac{p}{1 - qt}$$

where $q = 1 - p$.

(ii) Use $G(t)$ to find the mean and variance of X.

10 The independent random variables W, X and Y have Poisson distributions with parameters λ_1, λ_2 and λ_3, respectively.

(i) Write down expressions for $P(W = w), P(X = x)$ and $P(Y = y)$.

(ii) Write down the PGF of the random variables W, X and Y. Hence obtain the PGF of $Z = W + X + Y$.

(iii) Use this PGF to find

(a) $P(W + X + Y = k)$, where k is a non-negative integer

(b) the mean of $W + X + Y$

(c) the variance of $W + X + Y$.

✓ KEY POINTS

1 For a discrete probability distribution, the probability generating function (PGF) is defined by

$$G(t) = E(t^X) = \sum_x t^x p_x,$$

where $p_x = P(X = x)$.

2 The sum of probabilities $= 1 \rightarrow G(1) = 1$.

3 *Expectation:* $E(X) = G'(1) = \mu$.

4 *Variance:* $\mathrm{Var}(X) = G''(1) + G'(1) - (G'(1))^2 = G''(1) + \mu - \mu^2$.

5 Special probability generating functions

Probability distribution	P(X = r)	PGF $G_X(t)$	Expectation E(X)	Variance Var(X)
Binomial	$\binom{n}{r} p^r q^{n-r}$	$(q + pt)^n$	np	npq
Poisson	$e^{-\lambda} \dfrac{\lambda^r}{r!}$	$e^{\lambda(t-1)}$	λ	λ
Geometric	pq^{r-1}	$\dfrac{pt}{1 - qt}$	$\dfrac{1}{p}$	$\dfrac{q}{p^2}$

6 For two independent random variables X and Y

$$G_{X+Y}(t) = G_X(t) \times G_Y(t).$$

Answers to exercises are available at <u>www.hoddereducation.com/cambridgeextras</u>

LEARNING OUTCOMES

Now you have finished this chapter, you should be able to

- understand the concept of a probability generating function (PGF)

- construct and use the PGF for given distributions (including the discrete uniform, binomial, geometric and Poisson distributions)

- use formulae for the mean and variance of a discrete random variable in terms of its PGF

- use formulae for the mean and variance to calculate the mean and variance of a given probability distribution

- use the result that the PGF of the sum of independent random variables is the product of the PGFs of those random variables.

Index